AF886407

ExcelTips

Powerful Lookup Functions

Second Edition

Using Excel's Powerful Lookup Functions to Find and Access Just the Data You Need

tips.NET

ExcelTips: Powerful Lookup Functions

Published by:

Sharon Parq Associates, Inc.
PO Box 794
Orem, UT 84059

For information on purchasing, distributing, or reselling books published by Sharon Parq Associates, Inc., please visit our website (www.SharonParq.com) or call 801-607-2035. Our books are also available through select online resellers such as Amazon.

ISBN: 978-1-61359-079-9 (printed book)
ISBN: 978-1-61359-080-5 (e-book)
ISBN: 978-1-61359-081-2 (e-book on CD-ROM)

Produced and published in the United States of America

Revision history:

1 December 2009: First edition
29 February 2012: Second edition

Table of Contents

Introduction

Excel provides hundreds of worksheet functions that you can use to build formulas, modify data, and get just the information you need in a cell. Of all the worksheet functions available, few are more powerful, as a group, than the lookup functions. The functions in this group allow you to indirectly access information in a worksheet; they allow you to "look up" information based upon various parameters that you specify.

In total there are seven lookup functions provided in Excel:

- CHOOSE
- MATCH
- INDEX
- OFFSET
- VLOOKUP
- HLOOKUP
- LOOKUP

While the number of functions may seem small, the power these functions provide you is anything but small. The lookup functions are some of the most misunderstood worksheet functions in Excel. They are definitely not the first functions people think of when learning how to use Excel, but they are essential to understand if you want to master how your data is accessed in your formulas. Once you understand how to use these functions, you'll see numerous ways in which you can enhance your worksheets.

The pages of ***ExcelTips: Powerful Lookup Functions, second edition,*** focus on dozens of tips that can help you better understand not only what lookup functions are, but how to use them in real-world scenarios to improve the calculations you must do in your own workbooks. In short, ***ExcelTips: Powerful Lookup Functions*** provides a comprehensive collection of information not available in any other single source.

Many people consider Excel's lookup functions, by their nature, to be an advanced topic. Even so, the tips in this book are written so that even the most novice of users can understand how to use the functions. Indeed, there is plenty of information here for a broad range of readers.

Regardless of your experience level with Excel—and because any good exposition demands a good foundation—you'll want to make sure you read through the first major section in this e-book ("Lookup Function Basics"). It lays the groundwork that is needed to understand everything else in the rest of the e-book, and without it you won't get as far as quickly as you will with it.

A Word about Versions

This e-book is written in such a way that it is immediately usable, regardless of which version of Excel you are using. Lookup functions are used as building blocks in formulas, and formulas are generally constructed the same regardless of the version of Excel you are using. In other words, the information here should be usable in all modern versions of the program, from Excel 97 through Excel 2010.

Due to the different user interfaces used in various versions of Excel, there are some differences in how some tasks are accomplished. If a tip involves the use of the user interface, those differences are taken into account and differing instructions are provided. Keep your eyes peeled for those differences.

Macros and Array Formulas

Macros provide a way for you to expand what can be done with Excel. They are great for quickly doing mundane, repetitive tasks and they are wonderful at processing huge amounts of data. Array formulas allow you to access groups of data (arrays) in ways not possible with regular formulas.

Why do I bring this up? Because you will notice repeated references to both macros and array formulas in this book. Macros often provide ways to accomplish some tasks faster or more completely than can be done with lookup functions alone. The applicability of this statement varies from tip to tip, so you'll want to evaluate each instance on its own terms. Each tip that provides a macro (not all of them do) touches, in some way, on the use of lookup formulas, so they will easily contribute to your overall knowledge of the topic being discussed.

Array formulas, on the other hand, are a natural companion to lookup functions. You'll find many places where a task that uses a lookup function is accomplished by sticking that function in the middle of an array formula. Each such use is clearly called out throughout this e-book.

Remember that this book is a resource for lookup functions; it is not a resource for developing or using either macros or array formulas. If your interest is piqued and you want more information on putting these aspects of Excel to work, consider the following e-books:

- ***Microsoft Excel VBA Guidebook.*** Creating Excel macros allows you to extend your productivity with Excel. Visual Basic for Applications (VBA) is the programming language used for creating Excel macros. Here is your expert guide on how to finally master creating macros
- ***ExcelTips: The Macros.*** Want to know anything and everything about how to create and use macros in Excel? You'll find it in *ExcelTips: The Macros.* At almost 800 pages, this volume contains more information about Excel macros than you will find anywhere else on the planet.
- ***ExcelTips: Amazing Array Formulas.*** Once you understood how to use them, array formulas provide new and exciting ways that you can perform calculations with your data. In fact, there are some calculations that can be done with array formulas that would otherwise be impossible with regular Excel formulas. All this information is detailed completely in *ExcelTips: Amazing Array Formulas.*

More information on either of these e-books can be found at the Tips.Net Store (http://store.tips.net).

Need More Tips?

Remember that this document contains tips presented in ***past*** issues of ***ExcelTips***. If you want to stay at the top of your Excel form, you should consider subscribing to ***ExcelTips***. All you need is an e-mail account, and you can receive a weekly dose of tips related to Excel. To subscribe to ***ExcelTips*** visit the ***ExcelTips*** Web site:

http://excel.tips.net

If you prefer to only consider tips that work with Excel 2007 or Excel 2010, then you'll want to visit this site:

http://excelribbon.tips.net

ExcelTips is part of the Tips.Net network, a collection of websites designed to bring you helpful tips, tricks, and ideas that can save you time, effort, and money. You can see the full breadth of the Tips.Net network by visiting our website:

http://www.tips.net

Sharing this Document

There was a lot of work that went into putting this document together. I can't tell you how many countless hours are spent putting together the weekly ***ExcelTips*** newsletter from which much of this information was compiled. That means that this information has value, and your friends, neighbors, and co-workers may want to share it.

The information in this document is copyrighted. I would ask that you do not share this information with others—you purchased this book, and you have a right to use it on your system. Another person who has not purchased this book does not have that right. It is the sales of this valuable information that makes the continued publishing of ***ExcelTips*** possible. If enough people disregard that simple economic fact, the newsletter will no longer be viable or available.

If your friends think this information is valuable enough to ask you for it, they should think it is valuable enough to purchase on their own. After all, the price is low enough that just about anyone should be able to afford it.

It should go without saying that you cannot post this document or the information it contains on any electronic bulletin board, Web site, FTP site, newsgroup, or ... well, you get the idea. The only place from which this document should be available is the ***ExcelTips*** Web site. If you want an original copy, visit the Tips. Net store at the following address:

http://store.tips.net/

1

Lookup Function Basics

Understanding the CHOOSE Function

Perhaps the simplest lookup function provided by Excel is the CHOOSE function. Just because it is simple, however, does not mean that it cannot be used in powerful ways to enhance your formulas.

The idea behind CHOOSE is that it, well, "chooses" a particular value from a number of optional values. For example, let's say that you have a series of five choices and you want to choose the second of those choices:

```
=CHOOSE(2,"First","Second","Third","Fourth","Fifth")
```

In this example the first parameter (the number 2) specifies which option from those that follow should be selected. In this case the second option is the text string "Second", so that is what is returned by the formula.

Obviously this example has only limited value. The CHOOSE function becomes much more powerful when you replace the first parameter with a cell reference or with an expression that returns a value between 1 and either 254 (Excel 2007 or Excel 2010) or 29 (earlier versions of Excel). The expression or the cell reference must resolve to a value in this range because you can have only up to either 29 or 254 options in the CHOOSE function, depending on your version of Excel. If you provide a value that is less than 1 or greater than the upper limit, then CHOOSE returns a #VALUE! error.

The real power in CHOOSE comes when it is combined with other functions. In fact, you'll find numerous examples throughout this e-book where CHOOSE forms an integral part of the lookup formulas used.

Understanding the MATCH Function

The MATCH worksheet function is very versatile. It allows you to compare a value to a range of values and then, if the value is found in the range, return the position within the range at which the value was first located. For instance, if you have a value in cell B18 and you need to compare that value to whatever is in cells A18:A493 you could use the following:

```
=MATCH(B18,A18:A493,0)
```

You can probably deduce from the description that the first parameter is a cell reference to the value you want to find and the second parameter is the cell range to which the first parameter should be compared. What about that third parameter (0), however? What does it do?

The third parameter for MATCH can be one of three values, and the value determines how the comparison is done by the function:

- **1 or TRUE:** Find the position of the largest value that is less than or equal to the value of the first parameter.
- **0 or FALSE:** Find the position of the first value that is exactly equal to the value of the first parameter.
- **-1:** Find the position of the smallest value that is greater than or equal to the value of the first parameter.

If you decide to leave off the third parameter (it is optional), MATCH assumes a value of 1 or TRUE.

Which "match type" (as they are called by Microsoft) you select says something about the character of the array you specify in the second parameter. If you specify a match type of 1, then your data should be sorted in ascending order; specify -1 and your data should be sorted in descending order. In the case of a match type of 0 it doesn't really matter what order your data is in.

This need to sort your data based on the match type you are using really makes sense, if you think about it. If you want MATCH to find the largest value that is less than or equal to whatever you are searching for and if the data is in ascending order, then MATCH can stop looking through the range when it encounters a value that is larger than the one you are search for; this means that the function does its work much faster when working with the sorted data.

There is one other unique behavior of MATCH if you are using a match type of 0 and if your data is text. In this case, what you are search for can contain wildcard characters: either * (any sequence of characters) or ? (any single character). Thus, searching for "se*" would return the position of the first instance of any text beginning with "se".

It is important to remember that the MATCH function returns a position of a value, not the actual value itself. For instance, in the example used earlier in this tip if the MATCH function returns 7, that means that what you were searching for (the value in B18) was found at the seventh cell in the range (A18:A493), which is cell A24.

Because MATCH returns positional information rather than actual values, it is often used in conjunction with other functions that can use that information. You can find several instances of MATCH being used in that manner throughout this e-book.

Understanding the INDEX Function

The INDEX function is used to reference the value at a cell that is offset from the top-left cell in a range that you specify. It only requires three parameters, as shown here:

```
=INDEX(A4:D7,3,2)
```

The first parameter is the range in which the cell is to be looked up, the second parameter is how many rows offset from the top left (so 3 references row 6), and the third parameter is how many columns offset from the top left (so 2 references column B).

This may seem a strange way to reference cell B6, but it becomes very powerful when you combine INDEX with other lookup functions, particularly the MATCH function. Since you can use MATCH to return an offset within a range, then you can use that offset as one of the parameters for INDEX. Together they are a powerful combination, as evidenced by their use together in several places in this e-book.

Understanding the OFFSET Function

One of the strangest little worksheet functions you can find in Excel is the OFFSET function. (OK; it may not be the strangest, but it ranks right up there among a rather large group of strange functions.) It allows you to return either a range address or a value at a range address defined by five parameters:

- A starting point in the form of a cell address
- The number of rows to be offset from the starting point (positive moves down and negative moves up)
- The number of columns to be offset from the starting point (positive moves right and negative moves left)
- The number of rows in the desired range
- The number of columns in the desired range

Thus, the following formula would, for all intents and purposes, return the range address of B7:D42:

```
=OFFSET(A4,3,1,36,3)
```

I say "for all intents and purposes" because if you use the formula exactly as shown you will get the #VALUE! error. In this case an error value was returned because the function is trying to return a multi-cell range (B7:D42) into a single cell (whatever cell contains the formula). In cases like this OFFSET is really meant to be used with other functions. For instance, you could wrap the function in the SUM function, as shown here:

```
=SUM(OFFSET(A4,3,1,36,3))
```

This will not produce an error, but instead will provide the sum of all the cells in the range B7:D42. Granted, it would have been just as easy to provide the actual range (B7:D42) in the SUM function, but that misses the point—you can use variable values with OFFSET or create a range reference that is anything but static; it can be shaped by those values to whatever is appropriate for your needs.

For example, you may want to determine the minimum temperature in a data table that starts at cell A3 and can extend across an unknown number of rows and will have as many columns as there are days in the month so far. You could construct such a formula using the OFFSET function in this manner:

```
=MIN(OFFSET(A3,0,0,COUNT(A3:A99),DAY(TODAY())))
```

Every day this formula pulls from a different range because the number of columns is defined by whatever day of the month it happens to be.

It should be mentioned that the fourth and fifth parameters used with OFFSET (the number of rows and the number of columns) are optional. If you leave these two parameters out of the function, then the resulting range is the same height and width as the range specified for the first parameter. Thus, in the original example used in this tip, you could have left off the last two parameters and it would not have returned an error value:

```
=OFFSET(A4,3,1)
```

This returns the value of whatever is in cell B7; it doesn't return a range address. Again, the power behind using the function in this manner depends on what you use for the parameters.

Understanding the VLOOKUP Function

Excel has a huge number of different worksheet functions—some of them used quite often, and others quite obscure. One of the more obscure functions (at least for Excel novices) is the VLOOKUP function. Understanding how this function works can make your life much simpler, especially when dealing with tables of data and pulling information from those tables.

VLOOKUP is a shortened form of "Vertical Lookup." It is a function that looks vertically (up and down) through a data table and extracts information from the table as you direct. For instance, let's assume you have a data table that lists part numbers and their prices. The part numbers are in column H, and the prices for those parts are in column I. (Assume the data table is in the range H5:I27.) You can use the following formula to look up a part number (located in cell C28) and return its price:

```
=VLOOKUP(C28,H5:I27,2)
```

Notice that, at a minimum, VLOOKUP requires three arguments. The first is the value that you want to look up. In this case, C28 contains the part number to be matched in the data table. The second argument is the actual data table, in this case H5:I27. The third argument specifies from which column of the table the value should be returned. In this case I wanted the price, which was in the second column (column I) of the data table.

What VLOOKUP does, in this instance, is to take the value in C28 and then try to match it to a cell in the first column of the range H5:127. If it finds a match, then it returns the value from column 2 of that range—the price we wanted.

VLOOKUP also accepts an optional fourth argument, which can be either TRUE or FALSE. The default value for the argument is TRUE, which means that VLOOKUP will approximate values when matching them in the data table. If an exact match cannot be found in the data table, then VLOOKUP considers the next lowest value in the first column of the table to be a match. Thus, if you are looking for a part number such as "P23," and there is no such part number in the table, but there is a "P22," then VLOOKUP considers that a match. If you set the optional fourth argument to FALSE, then VLOOKUP only returns successfully if it can make an exact match.

Because of the way in which VLOOKUP does its matching, it is very important that the information in your data table be sorted in ascending order according to the values in the first column.

Understanding the HLOOKUP Function

The HLOOKUP function is very similar in purpose and function to the VLOOKUP function. The major difference can be discerned from the differences in the function names: Whereas VLOOKUP does its magic on a vertical table, HLOOKUP does its on a horizontal table.

HLOOKUP (short for "Horizontal Lookup") looks horizontally (left and right) through a data table and extracts information from the table as you direct. For instance, let's assume you have a data table that contains sales regions in row 3 (North, East, South, etc.) and sales quantities beginning in row 4. (Assume that the data table is in the range B3:Q57.) You can use the following formula to look up a sales region (from cell A2) in the table and return the sales quantity from the fifth row of the table:

```
=HLOOKUP(A2,B3:Q57,5)
```

The HLOOKUP function requires, at a minimum, three arguments. The first is the value that you want to look up (cell A2), the second is the actual data table (the range B3:Q57), and the third specifies from which row of the table the value should be returned (row 5).

What HLOOKUP does, in this instance, is to take the value in cell A2 and then try to match it to a cell in the first row of the range B3:Q57. If it finds a match, then it returns the value from the fifth row of that range.

HLOOKUP, just like VLOOKUP, also accepts an optional fourth argument that can be either TRUE (1) or FALSE (0). The default value for the argument is TRUE, which means that HLOOKUP will approximate values when matching them in the data table. If an exact match cannot be found in the data table, then HLOOKUP considers the next lowest value in the top row of the table to be a match. Thus, if you are looking for a sales region such as "Outside," and there is no such sales region in the table, but there is a "Midwest," then HLOOKUP considers that a match provided there is nothing "closer" to the desired text you are looking up. If you set the optional fourth argument to FALSE, then HLOOKUP only returns successfully if it can make an exact match.

Because of the way in which HLOOKUP does its matching, it is very important that the information in your data table be sorted in ascending order, left to right, according to the values in the first row of the table.

Understanding the LOOKUP Function

The LOOKUP function is, in many ways, a holdover from the earliest days of Excel. It is included in the family of lookup functions largely for compatibility with older worksheets or with information that originated in other spreadsheet programs, such as Lotus 1-2-3. In most instances its functionality has been eclipsed by the HLOOKUP and VLOOKUP functions.

The LOOKUP function relies upon both a source and target range of cells. Microsoft refers to these as a "lookup vector" and a "result vector," respectively, but I believe that such nomenclature is unnecessarily technical. Basically the function finds a value in the source range and then returns a value from the same position in the target range. Consider the following example:

```
=LOOKUP(B4,E7:E22,H7:H22)
```

The value in B4 is compared to the cells in the range E7:E22. LOOKUP expects that the values in the range will be in ascending order, and it will match to the largest value that is less than or equal to whatever is in B4. If, for instance, this value is in the fifth cell in the range (E11), then the value in the fifth cell of the target range (H7:H22, with the fifth cell being H11) is returned.

LOOKUP can work with either a vertical range or a horizontal range. It is important to remember two things about the ranges, however. First, the ranges can only be a single column wide or a single row high. Second, the target range must be the same size as the source range. It is perfectly permissible for the ranges to be different orientations (one can be horizontal and the other vertical), but they must contain the same number of cells each.

Maintaining Text Formatting in a Lookup

When you enter text into a worksheet, you can format that text in either of two ways: You can format all the text in the cell, or you can format just some of the text in the cell. Formatting all the text should be familiar to all readers, as it is quite common to make the contents of a cell bold or change its font.

Formatting some of the text within a cell may not be done as often, but it is just as easy to do as formatting the entire cell. All you need to do is, while editing the cell contents, select the text you want to format and then apply the desired formatting properties. For instance, you could make some portion of the text bold or some of it underlined. You can also change the font used by certain characters in a cell, which is often done to insert symbols in a cell or to use foreign-language characters within your regular text.

While formatting can make your information more understandable, and may in fact be required for your particular worksheet, there are some drawbacks. One of the biggest drawbacks is when you use formulaic references to the formatted text. Excel will only copy the bare contents of the referenced cell, not the formatting applied to text within that cell.

As an example, suppose you have a cell that contains Greek letters interspersed within the text of a cell, and you reference that cell in a formula such as the following:

```
=HLOOKUP(B7,A16:A29,C3)
```

Excel dutifully copies the contents of the cell, returning as the results of the formula, but it does not copy the formatting of the cell being referenced. This behavior is not limited to lookup functions, either; Excel behaves this way even with simple formulas, such as =B7.

There is no way around this drawback, short of writing a rather complicated macro that copies formats as well as content. Macros that copy cell formatting have been covered in past issues of ***ExcelTips;*** macros that copy individual character formatting within a cell are complex enough that they are beyond the scope of ***ExcelTips*** itself.

Errors when Copying References to External Cells

Kirk noted that if he copies a cell referencing a cell in an external workbook, he gets an error. However, if that external workbook is open when he does the very same steps, he does not get an error message. Kirk wonders why this occurs.

There could be a variety of reasons for this happening. The most probable is that for whatever reason the information that it must get from the external sheet is unavailable or cannot be located.

Another possible reason is that your formula could be an INDIRECT link or the workbook could have been opened and the links not updated. If the formula is complex, Excel requires that the external workbook be open. (Microsoft has never provided a clue as to what constitutes "complex" in this case. Basically, if you get an error, the formula is "complex.")

In general, you could consider a formula complex if it does a calculation on the external data to get a result, such as with MATCH, LOOKUP, etc. An example would be storing a month number in workbook A and using that value to calculate year-to-date costs from monthly data stored in individual cells in workbook B. (In other words, you use the month number as an index to access the information in workbook B.)

A workaround for some (but not all) situations is to do the complex calculations completely in workbook B with workbook B linking to workbook A to get the data needed for that calculation. Retrieving the month number from workbook A is not complex and can be easily done by Excel. You could then open workbook A and have it read the result of the calculation from workbook B.

2

Addresses and Cell References

Cell Address of a Maximum Value

Barry has a worksheet with 65,000 rows. They are unsorted and must remain unsorted. He can use the MAX function on the column and get the maximum value in that column. However, he also wants to know the address of the first cell in the column that contains this maximum value.

There are a number of ways that you can determine the address of the maximum value. One way is to use the ADDRESS function in conjunction with the MATCH and MAX functions, in the following manner:

```
=ADDRESS(MATCH(MAX(A:A),A:A,0),1,4)
```

The MATCH function is used to find where in the range (column A) the maximum value resides, and then the ADDRESS function returns the address of that location. A shorter version of the macro leaves off the ADDRESS function, instead being "hardwired" to return an address in column A:

```
="A"&MATCH(MAX(A:A),A:A,0)
```

Still another way to get the desired address is with a formula such as this:

```
=CELL("ADDRESS",INDEX(A:A,MATCH(MAX(A:A),A:A,0)))
```

This formula uses the CELL function, in conjunction with INDEX, to return the address of the cell that matches the maximum value in the column.

Finding the Address of the Lowest Value in a Range

When writing a macro, you can find the lowest value in a range of cells by using the WorksheetFunction method to apply the MIN worksheet function. You may need, however, to not only find the lowest value in the range, but also the address of the first cell that contains that value.

One simple way is to simply step through the range you want to examine and derive both the lowest value and the address of the cell being examined, as in the following:

```
Function FindLowestAddr(pRng As Range) As String
    Application.Volatile
    MinVal = pRng.Cells(1).Value
    MinAddr = pRng.Cells(1).Address
    For Each c in pRng
        If c.Value < MinVal Then
            MinVal = c.Value
            MinAddr = c.Address
        End If
    Next c
    FindLowestAddr = MinAddr
End Function
```

Note that this approach doesn't rely upon the MIN worksheet function at all. There is a drawback to it, however—it doesn't differentiate between cells that contain numeric values and those that don't. In other words, if the range passed to the function contains a blank cell, that cell is considered to contain a zero value, which may very well be the lowest value in the range.

One way around this is to rely upon worksheet functions from within the macro. The following macro uses both the MIN and MATCH worksheet functions to determine the location of the minimum value and then the index (offset) of that cell within the range.

```
Function GetAddr(rng As Range) As String
    Dim dMin As Double
    Dim lIndex As Long
    Dim sAddress As String

    Application.Volatile
    With Application.WorksheetFunction
        dMin = .Min(rng)
```

```
        lIndex = .Match(dMin, rng, 0)
    End With
    GetAddr = rng.Cells(lIndex).Address
End Function
```

It should be noted that if you are using the macro only to discover the address because you figured there was no way to derive the desired information without the macro, then you can do away with the macro entirely by using a worksheet formula. For instance, if you want to determine the address of the lowest-valued cell in the named range MyRange, you could use the following:

```
=ADDRESS(ROW(MyRange)+MATCH(MIN(MyRange),MyRange,0)-1,COLUMN(MyRange))
```

Referring to the Last Cell

There may be times when you are putting together a workbook and you want to keep a summary on one worksheet and detail information on another. For instance, let's suppose Sheet1 is your summary worksheet, and you have detailed information for bank accounts on Sheet2. In looking at the detail information, you have dates in column A, and balances for different accounts in columns B, C, and D. Thus, the detailed information is a table that shows a running progression of bank balances on different dates.

In putting together your summary information on Sheet1, you realize that you need to reference the last figures in columns B, C, and D. These figures represent the latest balances, and thus are perfect for your summary. How do you do it? Particularly when you continue to add information to your detail worksheet over time?

Actually, there are several ways to approach the problem. (There are typically several ways to solve any Excel problem.) One way is to use the VLOOKUP function. At the point in the summary where you want the latest balance from column B of the detail (Sheet2), you would put the following formula:

```
=VLOOKUP(MAX(Sheet2!$A:$A),Sheet2!$A:$D,2)
```

To change references for the other two account balances, you would simply change the last number (2) to either 3 (for the account in column C) or 4 (for the account in column D). The function works because it looks up the maximum value in column A, which contains dates. It then looks in the data table (Sheet2!$A:$D) and finds the appropriate offset for the desired column.

This approach works fine, provided there are no dates in column A past the last balances entered. If there are, then the values returned will always be incorrect.

Another way to approach the problem is to use the INDEX function in conjunction with either COUNT or COUNTA. If the detail columns don't contain any text (even in the column headers), then you would use the COUNT function. If there is text included, then COUNTA is preferred. At the point where you want to include the last balance from column B of the detail, you would use the following formula:

```
=INDEX(Sheet2!B:B,COUNTA(Sheet2!B:B))
```

It looks into the table, determines the number of non-blank cells in column B, and then pulls the figure from that last non-blank cell. To adapt the formula for columns C and D, simply change the B references to the appropriate C or D.

Still another way to deal with the problem is to use the OFFSET function, as in the following:

```
=OFFSET(Sheet2!B1,COUNTA(Sheet2!B:B)-1,0)
```

This function returns the value of a cell offset from a base reference cell. In this case, the base cell is Sheet2!B1. The COUNTA function is used to determine how many rows to offset from the base, and the 0 specifies that the offset should be in the same column as the base reference. To change the formula for columns C and D, simply change all references to B to either C or D.

Referencing the Last Cell in a Column

Patty asked about a common scenario, in which column B contains quite a bit of data, and information can be added to the cells in the column at any time. In a formula in cell C4, Patty wants to see the value at the bottom of those cells in column B that contain values. Thus, if cells B1:B27 contain data, then in cell C4 Patty wants to see the value that is in cell B27. If three more pieces of data are added to column B, then the value in C4 should contain the value in B30.

The solution to this problem depends on whether you can count on the data in column B containing blank cells or not. If the data is contiguous—it doesn't contain any blank cells—then you can use the following formula in C4:

```
=INDIRECT("B"&COUNTA(B:B))
```

This constructs an address based on the last cell in the column, and then uses the INDIRECT function to return the value at that address.

If it is possible for there to be blanks in column B, then the following formula will work:

```
=INDIRECT("B"&MAX(ROW(1:65535)*(B1:B65535<>"")))
```

Again, the INDIRECT function is used to fetch the actual value, but the address used by INDIRECT is put together differently.

A different approach is to use the VLOOKUP function to return the value. If column B consists of numeric values, then the following formula in C4 will work just fine:

```
=VLOOKUP(9.99999999999999E+307,B:B,1)
```

If column B contains text, then the numeric lookup won't work, but the following will:

```
=VLOOKUP(REPT("z",50),B:B,1)
```

Relative References within Named Ranges

Chris has set up a worksheet where he uses named ranges (rows) in his formulas. He has named the entire sales row as "Sales," and then uses the Sales name in various formulas. For instance, in any given column he can say **=Sales**, and the value of the Sales row, for that column, is returned by the formula. Chris was wondering how to use the same formula technique to refer to cells in different columns.

There are a couple of different ways this can be done. First of all, you can use the INDEX function to refer to the cells. The rigorous way to refer to the value of Sales in the same column is as follows:

```
=INDEX(Sales,1,COLUMN())
```

This works if the Sales named range really does refer to the entire row in the worksheet. If it does not (for instance, Sales may refer to cells C10:K10), then the following formula refers to the value of Sales in the same column in which the formula occurs:

```
=INDEX(Sales,1,COLUMN()-COLUMN(Sales)+1)
```

If you want to refer to a different column, then simply adjust the value that is added to the column designation in the INDEX function. For example, if you wanted to determine the difference between the sales for the current column and the sales in the previous column, then you would use the following:

```
=INDEX(Sales,1,COLUMN()-COLUMN(Sales)+1) - INDEX(Sales,1,COLUMN()-
COLUMN(Sales))
```

The "shorthand" version for this formula would be as follows:

```
=Sales - INDEX(Sales,1,COLUMN()-COLUMN(Sales))
```

There are other functions you can use besides INDEX (such as OFFSET), but the technique is still the same—you must find a way to refer to an offset from the present column.

There is an easier way to get at the desired data, however. Let's say that your Sales range also has a heading row about it, similar to what is shown in the following:

	D	E	F	G	H	I
8						
9	**2004**	**2005**	**2006**	**2007**	**2008**	**2009**
10	4301	4576	4455	4502	4528	4589

Example data for a worksheet

The heading row lists the years for the range, and the values under the headings are those that actually make up the Sales range. To make sure this technique will work, follow these steps if you are using Excel 2007 or Excel 2010:

1. Display the Excel Options dialog box. (In Excel 2007 click the Office button and then click Excel Options. In Excel 2010 display the File tab of the ribbon and then click Options.)
2. Click Formulas at the left side of the dialog box.

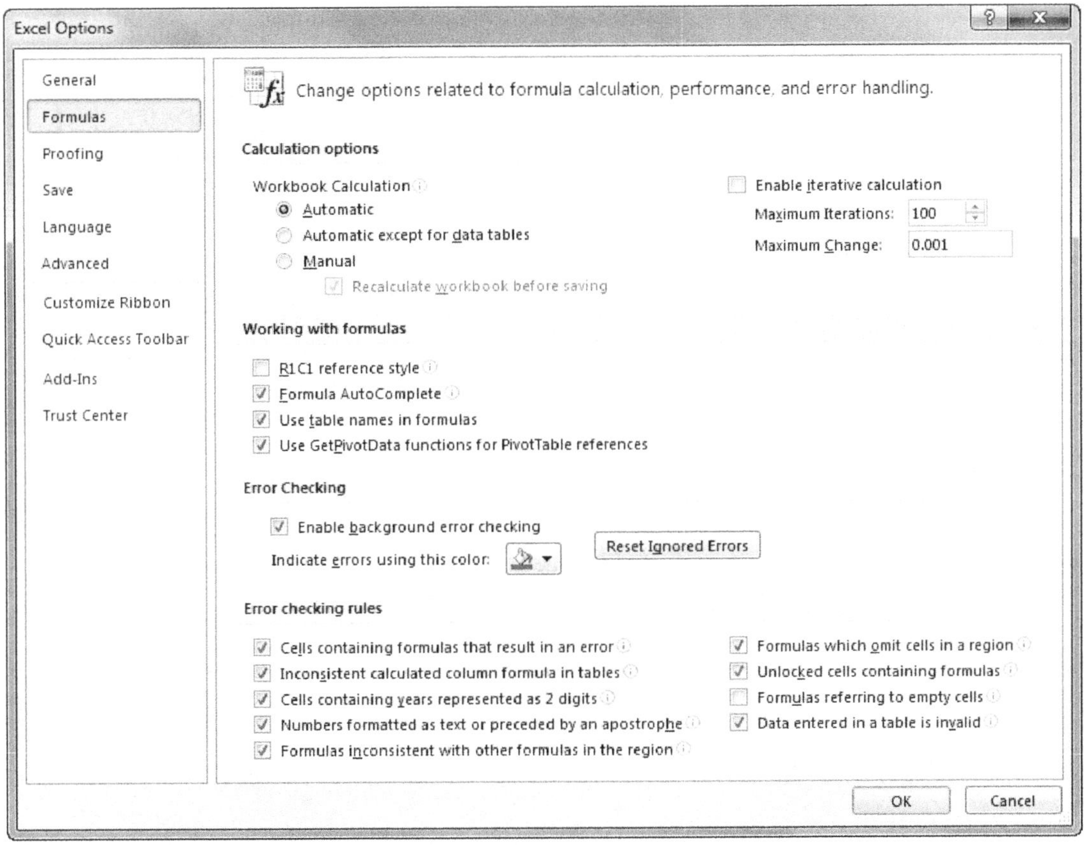

The Formulas options of the Excel Options dialog box.

3. Ensure that the Use Table Names in Formulas check box is selected.
4. Click OK.

For a version of Excel prior to Excel 2007, follow these steps instead:

1. Choose Options from the Tools menu. Excel displays the Options dialog box.
2. Make sure the Calculation tab is displayed.

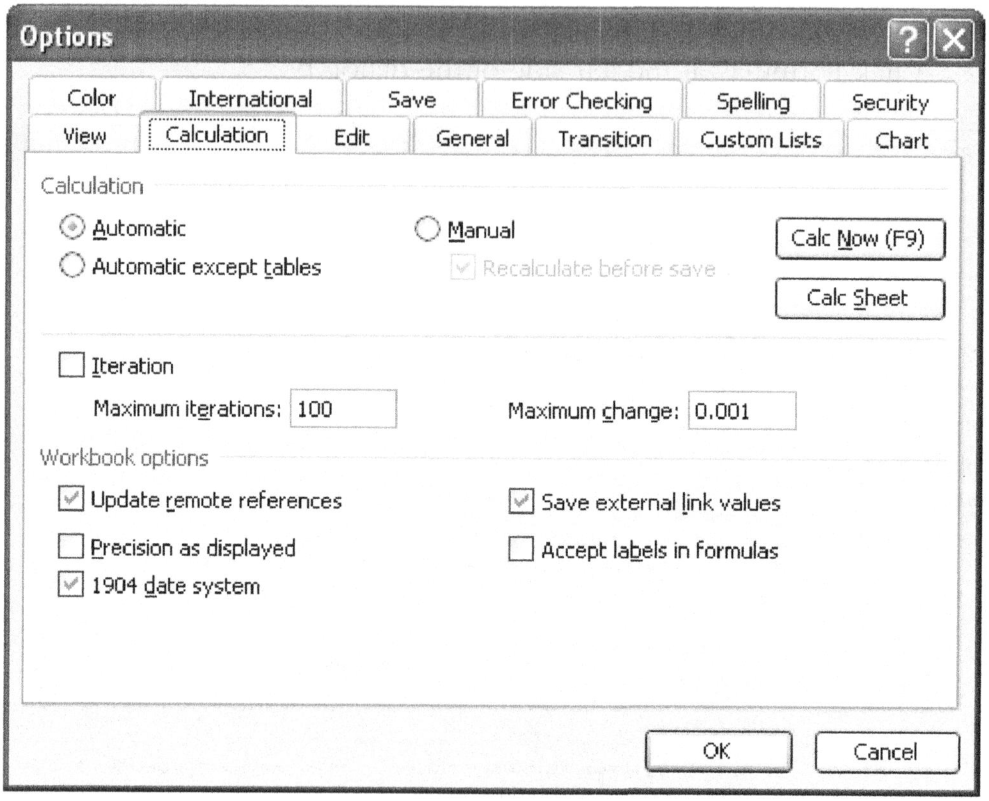

The Calculation tab of the Options dialog box.

3. Ensure that the Accept Labels in Formulas check box is selected.
4. Click OK.

With this configuration change done, you can use the following as your formulas:

```
=Sales '2009' - Sales '2004'
```

What you are actually doing is instructing Excel to work with unions of cells. In this instance, Sales '2009' returns the cell at the intersection of the Sales range and the '2009' column. A similar union is returned for the portion of the formula to the right of the minus sign. The result is the subtraction of the two values you wanted.

Non-adjusting References in Formulas

Everybody knows you can enter a formula in Excel. (What would a spreadsheet be without formulas, after all?) If you use address references in a formula, those references are automatically updated if you insert or delete cells, rows, or columns and those changes affect the address reference in some way. Consider, for example, the following simple formula:

```
=IF(A7=B7,"YES","NO")
```

If you insert a cell above B7, then Excel automatically adjusts the formula so that it appears like this:

```
=IF(A7=B8,"YES","NO")
```

What if you don't want Excel to adjust the formula, however? You might try adding some dollar signs to the address, but this only affects addresses in formulas that are later copied; it doesn't affect the formula itself if you insert or delete cells that affect the formula.

The best way to make the formula references "non-adjusting" is to modify the formula itself to use different worksheet functions. For instance, you could use this formula in cell C7:

```
=IF(INDIRECT("A"&ROW(C7))=INDIRECT("B"&ROW(C7)),"YES","NO")
```

What this formula does is to construct an address based on whatever cell the formula appears in. The ROW function returns the row number of the cell (C7 in this case, so the value 7 is returned) and then the INDIRECT function is used to reference the constructed address, such as A7 and B7. If you insert (or delete) cells above A7 or B7, the reference in cell C7 is not disturbed, as it just blithely constructs a brand new address.

Another approach is to use the OFFSET function to construct a similar type of reference:

```
=IF(OFFSET($A$1,ROW()-1,0)=OFFSET($B$1,ROW()-1,0),"YES","NO")
```

This formula simply looks at where it is (in column C) and compares the values in the cells that are to its left. This formula is similarly undisturbed if you happen to insert or delete cells in either column A or B.

A final approach (and perhaps the slickest one) is to use named formulas. This is a feature of Excel's naming capabilities that is rarely used by most people. Follow these steps if you are using Excel 2007 or Excel 2010:

1. Select cell C2.
2. Display the Formulas tab of the ribbon.
3. In the Defined Names group click the Define Name tool. Excel displays the New Name dialog box.

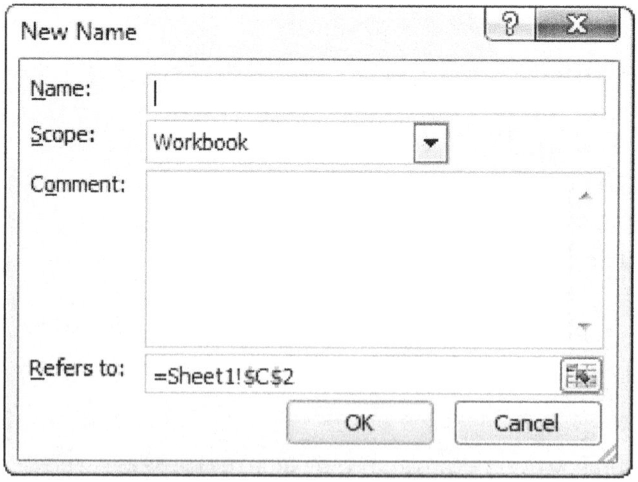

The New Name dialog box.

4. In the Name box enter the name **CompareMe**. (You can use a different name, if you desire.)
5. Erase whatever is in the Refers To box, replacing it with the following formula:

```
=IF(A2=B2,"YES","NO")
```

6. Click OK.

If you are using an older version of Excel, follow these steps instead:

1. Select cell C2.
2. Choose Name from the Insert menu, then choose Define from the submenu. Excel displays the Define Name dialog box.
3. In the Names in Workbook box, enter the name **CompareMe**. (You can use a different name, if you desire.)
4. Erase whatever is in the Refers To box, replacing it with the following formula:

```
=IF(A2=B2,"YES","NO")
```

5. Click OK.

At this point you've created your named formula. You can now use it in any cell in column C in this manner:

```
=CompareMe
```

It compares whatever is in the two cells to its left, just as your original formula was designed to do. Better still, the formula is not automatically adjusted as you insert or delete cells.

Stopping a Formula from Updating References

David wonders if there is a way he can make a formula not automatically update when he inserts a column that would otherwise affect the formula. For instance, if he has the formula =SUM(B:B) and then he inserts a column to the left of column B, the formula is automatically updated by Excel to =SUM(C:C). He doesn't want the formula to update; he still wants it to refer to column B after he inserts the new column.

One way to get the result you want is to use the OFFSET function to refer to column B. For instance, consider the following formula:

```
=SUM(OFFSET(A:A,0,1))
```

If this formula is in a cell, and you insert a column before column B, then the formula doesn't update; it still refers to column B. Why? Because the formula refers to column A and you didn't do anything to move column A. If you did insert a column before column A, then the formula would update to reference column B.

This means that the best way to handle the formula is to use the INDIRECT function, in this manner:

```
=SUM(INDIRECT("B:B"))
```

The INDIRECT function uses text for a parameter, and since it is text it is not considered a reference to be updated by Excel. Regardless of inserting or deleting columns, the formula will always refer to column B.

Incrementing References by Multiples when Copying Formulas

Tim has a worksheet in which cell B1 contains the formula =SUM(A1:A7). He wants to copy this formula down and have the range incremented by 7 rows, so that cell B2 would contain the formula =SUM(A8:A14), cell B3 would contain =SUM(A15:A21), etc. The problem is that when he copies it down, each "end" of the range is only incremented by 1 where it should be incremented by 7 to fulfill his need. He wonders how he can make Excel do the proper incrementing.

You can't make Excel do the proper incrementing using copy and paste; it just won't do it. The reason is simple—there are times when incrementing by 1 makes sense from a formulaic perspective. Since Excel can't read your mind (at least until the next version <g>), it makes the assumption that it should only increment by 1.

The solution is to change your formula. Using a couple of worksheet functions you can have Excel construct the desired range for the summation. Consider the following example of a formula that will provide the desired sum:

```
=SUM(INDIRECT("A" & (ROW()-1)*7+1 & ":A" & (ROW()-1)*7+7))
```

If you put this formula into cell B1, it works because it takes a look at the row number (returned by the ROW function) of the row in which the formula is contained. Since it is in row 1, then the formula is evaluated in this manner by Excel:

```
=SUM(INDIRECT("A" & (ROW()-1)*7+1 & ":A" & (ROW()-1)*7+7))
=SUM(INDIRECT("A" & (1-1)*7+1 & ":A" & (1-1)*7+7))
=SUM(INDIRECT("A" & 0*7+1 & ":A" & 0*7+7))
=SUM(INDIRECT("A" & 0+1 & ":A" & 0+7))
=SUM(INDIRECT("A" & 1 & ":A" & 7))
=SUM(INDIRECT("A1:A7"))
=SUM(A1:A7)
```

What you end up with in B1 is the sum you desired. (The INDIRECT function uses the value in the string as if it was a real range, which is what you want.) When you copy the formula down the column, as the row number increments the formula provides the proper increments of 7 on both ends of the range.

There are other variations on this technique that you can use. The only difference is that the variations use different worksheet functions to accomplish the same task. For instance, the following variation still uses the ROW function, but then ultimately relies on the OFFSET function to calculate the desired range:

```
=SUM(OFFSET(A1,((ROW()-1)*6),0):OFFSET(A7,((ROW()-1)*6),0))
```

A shorter approach that uses OFFSET is as follows:

```
=SUM(OFFSET($A$1,ROW()*7-7,0,7,1))
```

Regardless of the approach, you can probably tell that the idea is to come up with a formula that uses the row in which the formula appears in order to construct the range that you really want. Each of the examples so far assumes that you are starting in cell B1. If you want to start in cell B2, then you'll need to modify the formulas to account for whatever row you are starting on. To give you just an idea of how this works, if you were starting in cell B2, instead, the three formulas presented in this tip would be modified in the following ways:

```
=SUM(INDIRECT("A" & (ROW()-2)*7+2 & ":A" & (ROW()-2)*7+8))
=SUM(OFFSET(A2,((ROW()-2)*6),0):OFFSET(A8,((ROW()-2)*6),0))
=SUM(OFFSET($A$2,(ROW()-1)*7-7,0,7,1))
```

Start at a different location, and you'll need to make further modifications to the formula you choose to use.

Where Is that Text?

Jay needs to determine the cell in which a particular text value occurs. He knows he can use Excel's Find and Replace capabilities to manually determine the address of cells containing a text value, but he is looking for a formula to determine the address of the cells. He wonders if there is a way to search for piece of text within a range and have Excel return the address of the cell in which the text is found.

There are two things you can try. First, if you are looking for an exact match for cell contents, then you can use a formula. The basic formula is this:

```
=ADDRESS(MATCH(C2,A:A,0),1)
```

In this example, cell C2 contains the value you are looking for and column A is the range of cells being searched. The formula returns a result regardless of the capitalization of C2 or the values in column A. Thus, if C2 contains "apple", then the formula will match positively to cells that contain "apple," "Apple," or "APPLE." Indeed, any mix of capitalization will match.

This formula will not return an address for a cell that contains what you are searching for amidst other text. So if you are searching for "apple" (cell C2), it won't return the address of a cell that contains the phrase "apple crisp." You can modify this behavior, a bit, by adding wild card characters to the search cell. For instance, if you search for "*apple*" then the formula returns the address of a cell that contains "apple", even if it is preceded or followed by other characters.

It should be pointed out that this formula only returns the address of the first cell in the range which meets the criteria. If you actually want the addresses of all cells that meet the criteria, then you'll need to rely on a macro. The following is a good example:

```
Function FindMe(x As Range, y As String) As String
    Dim r As Range
    Dim sResults As String
    Dim sSearch As String

    Application.Volatile
    sSearch = LCase(y)
    For Each r In x
        If InStr(1, LCase(CStr(r.Value)), sSearch) > 0 Then
            sResults = sResults & r.Address & ", "
        End If
    Next r
    If Len(sResults) > 2 Then
        FindMe = Left(sResults, Len(sResults) - 2)
    Else
        FindMe = ""
    End If
End Function
```

You use the function by simply providing the range you want to search along with what you want to search for:

```
=FindMe(A:A, "apple")
```

If you use a large range (as in this example—all of column A), then don't be surprised if the function takes a noticeable amount of time to return a result. This makes sense, as it has to search through every cell in the range, regardless of whether there is anything in the cell or not.

You also don't need to use any wildcards with this function; it assumes that a match occurs if what you are looking for is located anywhere within the cell. It also doesn't pay attention to the capitalization of what you are looking for or the capitalization of anything in the search range.

3

Finding Duplicate Values or Unique Values

Synchronizing Lists

You may have an occasion when you have two data lists that we want to "line up." For instance, column A might be a customer account number, while column B displays the customer's account balance. In columns C and D you then paste a listing of customer payments, with column C being the customer account number and column D being the payment amount. Both lists (A/B and C/D) are sorted by customer account number.

Since not all customers with balances made payments, the A/B list is not in synch with the C/D list. To get them in synch, you need to insert blank cells where needed in columns C/D (and sometimes columns A/B) so that the customer account number in column C matches the customer account number in column A.

If your goal is to match payments to balances, then there is a relatively easy way to do this, without the need to insert cells in the lists. Follow these steps:

1. Insert three blank columns between the two lists. When done, you should have the account balances in A/B, blank columns in C/D/E, and the payments in F/G.
2. Assuming the first account/balance combination is in cells A2:B2, enter the following formula in cell C2:

```
=IF(ISNA(VLOOKUP(A2,F:G,2,FALSE)),0,VLOOKUP(A2,F:G,2,FALSE))
```

3. Copy the formula down through the rest of column C.

This formula looks in the payments columns (F/G) for any cells that match the account number in column A. If found, then the amount of the payment is returned by the formula. If a match is not located, then a zero value is returned.

The approach works well if you know that the payment columns contain only a single payment for each account. If it is possible that some accounts received multiple payments, then you need to change the formula you use in step 2:

```
=SUMIF(F:F, A2,G:G )
```

This formula, if it finds a match, adds all the payments together and returns the sum.

Of course, the example first described in this tip is just that—an example of a more pervasive problem. You may have a need to synch lists where there is only text in the lists, or where it is more difficult to do a lookup or you don't need to return a sum. In those instances, it may be best to look for a third-party solution.

Comparing Lists for Duplicates

Imagine for a moment that you have a worksheet that contains lists of part numbers. On one worksheet you have a list of part numbers, and on another worksheet you have a similar list. The lists are not identical, however, and you want to determine if a particular part number on one list also appears on the other.

One solution is to somehow combine the lists, but add some sort of indicator as to which original list the particular part number came from. This approach (or a variation thereon) is, in fact, the approach taken by many Excel users.

What if you don't want to combine the lists, however? In this case, there is a very easy way to do the comparison. Follow these steps:

1. Make sure there is a blank column just to the right of each list of part numbers on each worksheet.
2. Select the part numbers on the first worksheet and give them a name such as "PartList1". (In Excel 2007 or Excel 2010 display the Formulas tab of the ribbon and then click on Define Name in the Defined Names group. In older versions of Excel use Insert | Name | Define.)
3. Select the part numbers on the second worksheet and give them a name such as "PartList2".
4. Assuming that the first part number on the first worksheet is in cell A2, enter the following formula in cell B2:

```
=ISNUMBER(MATCH(A2,PartList2,0))
```

5. Copy the formula down so that a copy appears to the right of each part number on the first worksheet.
6. Repeat steps 4 and 5 on the second worksheet, but use the following formula:

```
=ISNUMBER(MATCH(A2,PartList1,0))
```

When you are done, either TRUE or FALSE will appear to the right of each part number on each worksheet. If TRUE appears, the associated part number appears on the other worksheet. If FALSE appears, then the part number is unique and does not appear on the other worksheet.

Another approach is to use an array formula to do the comparisons. You could follow the same steps shown above, but use the following formula in step 4 (and PartList1 variation in step 6):

```
=OR(EXACT(A2,PartList2))
```

Since this is an array formula, you would enter it by using **Shift+Ctrl+Enter**. The result is the same TRUE and FALSE designation described above.

Regardless of which formula approach you use, you can use the AutoFilter capabilities of Excel to limit what is shown on either worksheet. If you filter to show only the FALSEs, you will have a list of all unique part numbers. If you filter to show TRUEs, then you will have a list of duplicates.

Cleaning Up Lists

Larry works for a firm that does outbound calls to potential customers. The firm has downloaded the national Do-Not-Call list, and wants to check their targeted phone numbers against the list, so they can make sure they don't call anyone that is on the DNC list. He was wondering how this can be done in Excel.

First of all, the natural question is whether Excel is even the proper tool to use for such a task. In checking information at the Federal Trade Commission's Web site, it appears that the DNC list, which is a flat text file, can consist of either phone numbers in individual area codes or a file that contains all area codes. Quick calculations indicate that the average area code list has in excess of 300,000 phone numbers, with over 120,000,000 phone numbers nationwide.

Working with such large quantities of phone numbers in Excel is not only impractical, but virtually impossible—Excel can't handle that much data. A better solution would be to use some sort of database program (perhaps Access), which can work with much larger numbers of records. You could also search the Web for proprietary solutions that will work with the DNC list.

Assuming that you work with just a subset of the DNC list, and that it will all fit within your copy of Excel, then it is a relatively easy task to compare one list against another. This assumes that the data in your DNC list and the "need to check" list are in the same text format. For the sake of this example, assume as well that the DNC list is in column A, and the "need to check" list is in column C. You can then follow these steps:

1. Select the cell in column D that is just to the right of the first phone number you want to check in column C.
2. Enter the following formula in the selected cell:

```
=ISNA(MATCH(C2,$A:$A,0))
```

3. Copy the formula down so it is beside all the numbers you want to check in column C.

The results of the formula indicate whether the adjoining phone number is in the DNC list or not. If the result is TRUE, then the phone number is ***not*** in the DNC list; a result of FALSE means it is in the DNC list and should not be called. At this point you can easily sort the "need to check" list according to the results of the formula in column D. You can then delete all the phone numbers for which the value in column D is FALSE.

Ensuring Unique Values in a Column

When you are entering information in a worksheet, you may want to ensure that the data being entered is actually unique for a particular column. For instance, if you are entering a series of invoice numbers in column D of a worksheet, you may want to ensure that you don't enter the same invoice number twice.

There are a couple of ways that you can approach such a problem. If you are always entering your new information at the bottom of a column, you can use Excel's data validation feature to make sure that whatever you are entering in the current cell does not match anything higher up in the column. Simply follow these steps:

1. Select column D.
2. Display the Data Validation dialog box. (In versions of Excel prior to Excel 2007, choose Data | Validation. In Excel 2007 or Excel 2010 click the Data tab of the ribbon and then click Data Validation in the Data Tools group.)

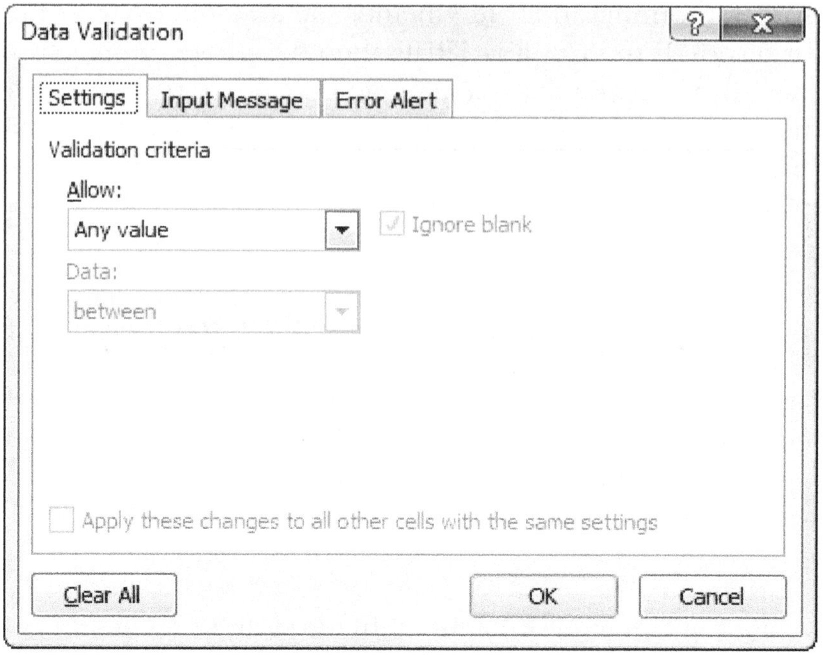

The Data Validation dialog box.

3. Using the Allow drop-down list, choose the Custom option.
4. In the Formula box, enter the following formula:

```
=MATCH(D1,$D:$D,0)=ROW(D1)
```

5. Click on OK.

If you may be entering data anyplace within column D, and you want to know if you are entering a value that is elsewhere in the column (above or below the current cell), then you can follow the above steps, but use this formula in step 4:

```
=COUNTIF($D:$D,D1)<2
```

Counting Unique Values

David has a worksheet in which there is a list of countries. This list, in cells A1:A100, can contain duplicates. David wants to determine the number of unique countries in the list.

There are several ways you can go about deriving a count, without resorting to using a macro. The method you should use depends on the characteristics of the data in the list. A good place to start, however, is to define a named range that represents the list of countries. In the following examples, it is assumed that the range is named Countries. (Catchy name, huh?)

If the list contains only text entries, and does not contain any blank cells, then the following will provide a count:

```
=SUM(1/COUNTIF(Countries,Countries))
```

This should be entered as an array formula, by pressing **Ctrl+Shift+Enter**. If the list contains blank cells, then the formula becomes a bit more complex. The following long array formula will work if there are blanks:

```
=SUM(IF(FREQUENCY(IF(LEN(Countries)>0,MATCH
(Countries,Countries,0),""), IF(LEN(Countries)
>0,MATCH(Countries,Countries,0),""))>0,1))
```

This approach—using the FREQUENCY function—is fully recounted in the Microsoft Knowledge Base:

```
http://support.microsoft.com/?kbid=100122
```

If you prefer to not use array formulas (for whatever reason), you can utilize a blank column to the right of your list. This column will contain regular formulas that indicate if the value to its left is unique in the list or not. The first time a value appears, the formula returns the number 1. On each subsequent appearance of the same value, the formula returns a 0. Start by sorting your list, then place the following formula in cell B1:

```
=IF(ISNA(VLOOKUP(A2,$A$1:A1,1,)),1,0)
```

Just copy the formula from B1 to the range B2:B100. With these results in place, you can easily sum column B and have a count of the unique values in the list.

Getting a Count of Unique Names

John has a worksheet that he uses for registration of attendees at a conference. Column A has a list of each person attending, and column B has the company represented by each attendee. Each company can have multiple people attend. John can easily figure out how many individuals are coming to the conference; it is simply the number of rows in column A (minus any header rows). The more difficult task is to determine how many companies are going to be represented at the conference.

There are a couple of ways to determine the desired count. First, if there are no blank cells in column B, you can use an array formula (entered by **Ctrl+Shift+Enter**) such as the following:

```
=SUM(1/COUNTIF(B2:B50,B2:B50))
```

If there are blanks in the range (B2:B50 in this case), then the array formula will return a #DIV/0! error. If that case, the array formula needs to be changed to the following:

```
=SUM(IF(FREQUENCY(IF(LEN(B2:B50)>0,MATCH(B2:B50,B1:B50,0),
""),IF(LEN(B2:B50)>0,MATCH(B2:B50,B2:B50,0),""))>0,1))
```

If you prefer to not use an array formula, you can add regular formulas to column C to do the count. First, sort the table of data by the contents of column B. That way the data will be in company order. Then add a formula such as the following to cell C2 (assuming you have a header in row 1):

```
=IF(B2<>B3,1,0)
```

Copy the formula down through all the rest of the cells in column C, and then do a sum on the column. The sum represents the number of unique companies attending, since a 1 only appears in column C when the companies change.

Of course, if you need to find the names of all the companies represented at the conference, you can use Excel's filtering capabilities. Follow these steps:

1. Sort the data by column B, the company names.
2. Select all the cells containing data (including the header cell) in column B.

3. Choose Filter from the Data menu, then choose Advanced Filter. Excel displays the Advanced Filter dialog box. (If you are using Excel 2007 or Excel 2010, display the Advanced Filter dialog box by displaying the Data tab of the ribbon and then, within the Sort & Filter group, clicking Advanced Filter.)

The Advanced Filter dialog box.

4. Make sure the Copy to Another Location radio button is selected.
5. Make sure the Unique Records Only check box is selected.
6. With the insertion point in the Copy To box, click on a blank cell, such as E1. (This is where the list of companies will be copied to.)
7. Click OK. Excel copies the unique company names from the original list to column E.

You now can easily see how many companies are being represented, along with who those companies are.

Finding Differences Between Lists

It is not unusual for people to keep client information in Excel worksheets. If you have a worksheet that contains the names of all your clients, and another worksheet that contains the names of your active clients, you may want to use Excel's capabilities to discover who your inactive clients are.

There are several ways you can accomplish this task. The first is through the use of VLOOKUP. This worksheet function works great, provided your lists of clients are arranged in alphabetical order. One way to use the function is to add a status column to your "all clients" worksheet. First, make sure you select your active clients and name them "Active". (How you define a name for a selected range of cells is covered in other ***ExcelTips***.) Then, in your full list of clients, add a column (named Status) to the right of your existing data. In the cells of the Status column, use the following formula:

```
=IF(ISNA(VLOOKUP(A2,Active,1,FALSE)),"Inactive","Active")
```

This formula assumes that the client's name is in column A of the current worksheet. The result of the formula is either "Active" or "Inactive," depending on whether there is a match between the name at A2 and the names in the Active list.

Once the Status column is in place, you can use the AutoFilter capability of Excel to filter your list based on the status column. You can then easily display the inactive clients, as desired.

It should be noted that while the above example uses the VLOOKUP worksheet function, you could just as easily compose other formulas that use functions such as HLOOKUP and MATCH. Which you use depends on your personal preferences and the way in which your data is laid out.

Another solution is to use a macro to compare each name on the "all clients" list with the names on the "active clients" list. If no match is found, then the name can be safely added to the "inactive clients" list. The following macro does just that:

```
Sub ListInactive()
    Dim cell As Range
    Dim SearchRng As Range

    Set SearchRng = Worksheets("Sheet2").Range("A:A")
    Counter = 1 'First row on Sheet3 contains headings
```

```
    For Each cell In Worksheets("Sheet1")
      .Range("A2:A1000") _
      .SpecialCells(xlCellTypeConstants)
        ID = cell 'Client ID
        NM = cell.Offset(0, 1) 'Client name
        MatchRow = 0
        On Error Resume Next
        MatchRow = WorksheetFunction.Match(ID, _
          SearchRng, 0)
        On Error GoTo 0
        If MatchRow = 0 Then
            Counter = Counter + 1
            Worksheets("Sheet3").Cells(Counter, 1) = ID
            Worksheets("Sheet3").Cells(Counter, 2) = NM
        End If
    Next cell
End Sub
```

The macro makes several assumptions about the data being examined. First, it assumes that the "all clients" worksheet is the first worksheet, and that the "active clients" worksheet is the second. Also, it is assumed that the third worksheet is blank and will end up containing the list of inactive clients. Further, the assumption is that column A contains a unique client ID number and column B contains the name of the client. When the macro is finished, the third worksheet contains the client numbers and names of all the inactive clients.

Getting Rid of Everything Except Numbers

Linda has a column that contains alpha and numeric characters. She needs to retain the numeric characters and delete the alpha ones. For example, a cell may contain 10003E111 and she wants to end up with 10003111.

There are a few ways you can approach this problem. Before proceeding with any solution, however, you should make sure that you aren't trying to change something that isn't really broken. For instance, you'll want to make sure that the "E" that appears in the number isn't part of the format of the number—in other words, a designation of exponentiation. If it is, then you don't really want to remove the character because it will end up changing the nature of the underlying number.

If you determine that the characters aren't part of the number's format, then you can first try using formulas to remove the alpha characters. If the values you want to change are in column A, you could enter the following (very long) formula in column B:

```
=MID(A1,MATCH(TRUE,ISERROR(1*MID(A1,ROW(INDIRECT
("1:"&LEN(A1))),1)),0),-MATCH(TRUE,ISERROR(1*MID
(A1,ABS(ROW(INDIRECT("1:"&LEN(A1)))-LEN(A1)-1),1))
,0)+LEN(A1)+2-MATCH(TRUE,ISERROR(1*MID(A1,ROW
(INDIRECT("1:"&LEN(A1))),1)),0))
```

Make sure you enter this as an array formula by pressing **Ctrl+Shift+Enter**. Then enter the following into column C:

```
=SUBSTITUTE(A1,B1,"")
```

The result is that column C contains the values from column A, without the alpha characters. You could use Paste Special to copy the information from column C to another column so that you end up with actual values instead of formula results.

This approach may work great for short-term use on a single workbook, but if you need to do this sort of data processing more often then you will want to create a user-defined function to do the processing. Here's an example:

```
Function OnlyNums(sWord As String)
    Dim sChar As String
    Dim x As Integer
    Dim sTemp As String

    sTemp = ""
    For x = 1 To Len(sWord)
        sChar = Mid(sWord, x, 1)
        If Asc(sChar) >= 48 And _
          Asc(sChar) <= 57 Then
            sTemp = sTemp & sChar
        End If
    Next
    OnlyNums = Val(sTemp)
End Function
```

You use this function by calling it from within a worksheet cell:

```
=OnlyNums(A1)
```

The function returns a numeric value. If you want to create an even shorter macro to do the processing, consider the following:

```
Function StripChar(aText As String)
    Dim I As Integer

    StripChar = ""
    For I = 1 To Len(aText)
        aChar = Mid(aText, I, 1)
        Select Case aChar
            Case "0" To "9"
                StripChar = StripChar & aChar
        End Select
    Next
End Function
```

To use this function, use either of the following in your worksheet:

```
=STRIPCHAR(A1)
=VALUE(STRIPCHAR(A1))
```

The first returns a text string consisting of the digits, the second returns the numeric version of that string.

Deleting All Names but a Few

Do you routinely work with worksheets that contain dozens (or hundreds) of named cells, and most of those names are unnecessary? Cleaning up the names is a huge task, but getting rid of the ones you don't need can make your workbook smaller and more efficient. The problem is, how do you get rid of a lot of unnecessary names all at once? You can certainly delete them one at a time, but such a process quickly gets tiresome.

One possible solution is to simply create a new workbook and copy the cells from the old workbook to the new one. Highlight the cells in the old workbook, use **Ctrl+C** to copy them, then paste them into worksheets in the new workbook. This will copy almost everything from the old workbook—formulas, formatting, etc. It does not bring copy over print settings or range names. The only task then remaining is to redefine the few names you want in the new workbook.

If you prefer to work with the old workbook (the one with all the names), it is best to create a macro that will do the name deletion for you. You need a macro that will allow you to delete all the names except those you want to keep. The following is a simple approach that accomplishes this task:

```
Sub DeleteSomeNames()
    Dim vKeep
    Dim nm As Name
    Dim x As Integer
    Dim AWF As WorksheetFunction

    'Add Names to keep here
    vKeep = Array("Name1", "Name2")

    Set AWF = Application.WorksheetFunction
    For Each nm In ActiveWorkbook.Names
        x = 0
        On Error Resume Next
        x = AWF.Match(nm.Name, vKeep, 0)
        On Error GoTo 0
        If x = 0 Then
            nm.Delete
        End If
    Next
    Set AWF = Nothing
End Sub
```

Before using the macro, modify the line that creates the vKeep array. Simply enter the names you want to keep within the array, each name surrounded by quotes and separated by commas. (In the example shown here, the names "Name1" and "Name2" will be kept.) The macro loops through all the names in the workbook and uses the MATCH function to see if the name is one in the array. If it is not, then it is deleted.

If you prefer to use a third-party solution to managing the names in your workbook, a great choice is the Name Manager add-in, written by Jan Karel Pieterse. You can find more information on the add-in here:

```
http://www.jkp-ads.com/officemarketplacenm-en.asp
```

4

Grabbing Info from a List

Returning a Weight and a Date

David asked how he could return a date associated with the maximum value in a particular column. In David's application, he had two columns containing dates and weights associated with those dates. He could figure out how to determine the maximum weight, but not how to pull the date on which that weight occurred.

In this instance, let's assume that the dates are in column A and the weights are in column B, rows 2 through 45. The following is the method of determining the maximum weight in column B:

```
=MAX(B2:B45)
```

You are not limited to using the MAX function; you can also use the LARGE function, which returns the largest specified value in a range. If you want the largest value, you use the function in this way:

```
=LARGE(B1:B45,1)
```

So far so good. In order to pull the corresponding date from the cell to the left of the date, all you need to do is use one of the lookup functions. The following example uses the INDEX function:

```
=INDEX(A2:A45,MATCH(MAX(B2:B45),B2:B45,FALSE),1)
```

The INDEX function pulls the value from row n, column 1 of range A2:A45. The value of n is determined by using MATCH to get the position number of the maximum value of range B2:B45 within that range. For example, if the highest weight is the fourth one on the list (the weight in row 5), the MATCH expression returns 4, therefore the INDEX function returns the date from row 5, column 1 of range A2:A45.

If you wanted, you could simplify the formula even more by just specifying column information within it. In this way, you could have as many entries as desired in columns A and B:

```
=INDEX(A:A,MATCH(MAX(B:B),B:B,FALSE),1)
```

You could just as easily use the LARGE function in place of the MAX function in these formulas. If there are two dates that have the same weight associated with them, then only the first matching date is returned. In addition, you will need to format whatever cell contains your formula with a date format.

If your columns are reversed, meaning that the weights are in column A and the dates are in column B, then you would use the VLOOKUP function instead of INDEX, as shown here:

```
=VLOOKUP(LARGE(A1:A45,1),A1:B45,2,FALSE)
```

Again, the MAX function could be used in this formula instead of the LARGE function, if desired.

Retrieving the Last Value in a Column

You may wonder if there is a way to return the last (not largest) value in a column. For instance, if there are values in A1 through A5, then you may want the value in A5 returned. Later, if values were added in A6 through A8, then the value in A8 should be returned.

There are a couple of ways that a solution can be approached. The first is to use a formula such as the following:

```
=INDEX(A:A,COUNT(A:A))
```

This formula returns the last numeric value in a column, providing that the values begin at (in this case) A1. This approach only works if all the values in the column are numeric. If the values are non-numeric, or if there are blank cells intermixed with the values, then a different approach is necessary. One way is to copy the following formula into column B, just to the right of the cells that may contain values:

```
=IF(ISNUMBER(A2),IF(A2<>0,ROW(A2),""),"")
```

In this case, the formula returns the row number of any cell in A which contains a numeric value greater than zero. The following formula can then be used to retrieve the last value in column A:

```
=INDEX(A:A,MAX(B:B))
```

This formula works because it returns the largest row number from column B, and then uses that as an index to return the corresponding value from column A.

As you can tell, returning the last value in a column can get a bit tricky at times. A clean approach is to simply develop your own VBA function that returns the desired value. In this case you can program the function to return any value—not just numeric values. A fine example of such a function is available at John Walkenbach's Web site. Check out the following:

```
http://j-walk.com/ss/excel/tips/tip30.htm
```

Last Non-Zero Value in a Row

Brian has a row of numbers with 240 cells. In this row, the numbers are steadily declining and will eventually, at some point in those 240 cells, become 0. The zeroes will continue to fill the remaining cells in the row. Brian needs to write an equation that will return the last non-zero value in the row.

There are a variety of ways that the desired value can be returned. (Doesn't that always seem to be the case with Excel? You can come up with lots of ways to get a result.) In general, you could use a regular formula or an array formula.

If you want to use a regular formula, here's one you can try:

```
=OFFSET(A6,0,(COUNT(A6:IF6)-COUNTIF(A6:IF6,0))-1)
```

The COUNTIF function counts the number of zero values and the COUNT function determines the number of cells in the range. Subtracting one from the other and adjusting by 1 gives the OFFSET value into the "array" of cells where the last non-zero value lies. This formula assumes that the values begin in column A; if they begin in a different column then you'll need to adjust the value provided by the COUNT/COUNTIF portion of the formula to represent the offset from the first column.

Here's a shorter variation of the formula, based on doing an offset from the right side of the range rather than the left side:

```
=OFFSET(IF6,0,-COUNTIF(A6:IF6,0))
```

In this instance it is important that IF6 be the actual right end of the range. The formula works by counting the number of zero values in the range (all at the right side of the range) and then computing the cell address of the last cell (IF6) minus the number of zeros.

Here is a version that uses the INDEX function, instead:

```
=INDEX(A6:IF6,,MATCH(0,A6:IF6,0)-1)
```

This version is even shorter, using the LOOKUP function:

```
=LOOKUP(1,1/(6:6>0),6:6)
```

Array formulas can also be used. (Array formulas are entered by pressing **Ctrl+Shift+Enter**.) This one uses the INDIRECT function:

```
=INDIRECT("R6C" & MAX((A6:IF6>0)*COLUMN(A6:IF6)),FALSE)
```

This array formula uses an interesting implementation of the LOOKUP function to find the correct result:

```
=LOOKUP(9.99999999999999E+307,IF(A6:IF6<>0,A6:IF6))
```

Here's another array formula that can be used, this time using the OFFSET function to find the last non-zero value in row 6:

```
=OFFSET(A6,0,MIN(IF(6:6=0,COLUMN(6:6),300))-2)
```

Here's an even shorter variation:

```
=MIN(IF(A6:IF6>0,A6:IF6))
```

All of these formulas presented so far depend on the fact that the numbers in the row actually do decline—they go from whatever the beginning number is and steadily go toward zero. If the numbers don't decline, then you can use a different type of array formula to determine the last non-zero value in the row:

```
=INDEX(6:6,MAX(IF(A6:IF6<>0,COLUMN(A6:IF6))))
```

The formula first determines the maximum column in the row (in this case row 6) that has a value not equal to zero, then it uses the INDEX function to get the value from that column in that row.

As you can tell, there are quite a few ways to find the last non-zero value in a row. Pick the one that strikes your fancy; there is no right or wrong in this instance.

Determining a State from an Area Code

Steve has phone numbers in column A, such as 3035551212, and would like to be able to look at the first three digits (the area code) and return in column B the state with which that area code is associated. He wonders about the best way to accomplish this.

Excel has many functions that make life easier when you are trying to manipulate data. In this case, using the VLOOKUP function to match the area code to the corresponding state is simple.

Before applying a function to retrieve the information you want, you need to create a simple data table that contains the data you want retrieved. In this table, you need to have the area codes and states in their own columns side-by-side, within your worksheet, sorted by area code. For instance, you might put the area codes in column F and the states for those area codes in column G. The area codes and states can be found on a number of websites, or you can create your own table if you prefer.

Once you've got the data into the two columns, select those columns and create a name for the selected range. (How you create a named range has been covered in other issues of ***ExcelTips***.) For example, you might name the range something like StateCodes. This naming, while not strictly necessary, makes using the lookup formula easier.

Assuming that the phone number is in cell A1 and that you would like the state name returned in the column next to the phone number, in cell B1 you would enter:

```
=VLOOKUP(VALUE(LEFT(A1,3)),StateCodes,2,FALSE)
```

The VALUE and LEFT functions are used to pull just the first three characters from the phone number. This is then used in the VLOOKUP formula to find the area code in the StateCodes table. Excel returns the name of the state than corresponds with the area code.

Another way that you can pull out the area code (which is essential for the lookup) is to use the FLOOR function, as shown here:

```
=VLOOKUP(FLOOR(A1/10000000,1),StateCodes,2,FALSE)
```

Note that this approach requires that the phone number be stored as a number, so it can be divided by 10,000,000.

The approaches discussed here work great, provided that your phone numbers are always in the specified format (3035551212). If your phone numbers have a different format—perhaps one that uses parentheses and dashes—they the formula won't work and will need to be adjusted to look at where the area code really is in the phone number. If you have phone numbers that are not in a single format, then all bets are off and the task of doing the lookup becomes much, much harder.

Figuring Out the Low-Score Winner

Mike asked if there is a way to determine the low-score winner on a hole-by-hole basis in a golf game. He uses Excel to track the individual hole scores, but he needs to know who has the lowest unique score on each hole. (If there is a tie on a hole, then he doesn't need to know who was involved in the tie.)

In providing an answer, there are several assumptions that must be made. First, assume that the values 1-18 (for each golf hole) are in cells A2 through A19. Second, assume that the low-score winner will be noted in column B. Third, assume that there are four golfers playing, and that their names are in cells C1 through F1. This range (C1:F1) is named "GolferNames". Finally, the golf scores for each golfer are entered in cells C2 through F19.

With this structure used, there are any number of ways that the formula could be put together. I particularly like this formula, which should be placed in cell B2:

```
=IF(MIN(C2:F2)=SMALL(C2:F2,2),"There is a " &
TEXT(COUNTIF(C2:F2, MIN(C2:F2)),"0") & "-way tie",
INDEX(GolferNames,MATCH(MIN(C2:F2),C2:F2,0)))
```

This is a very long formula, and you should make sure that it is entered all on a single line. You can then copy the formula from B2 and paste it in B3 through B19.

If there is a tie (determined by comparing the results of the MIN function with the second lowest score, as returned by the SMALL function), then the formula returns "There is a 2-way tie", or whatever number is actually involved in a tie. If there is not a tie, then the INDEX function is used to retrieve the name of the golfer that had the lowest score for the hole.

This example used, of course, only four golfers. If there are more golfers involved, the only alterations to make involve changing the range covered by the GolferNames range and expanding all instances of C2:F2 in the formula to represent the actual range of golfer scores.

If you prefer to simply not list anything if there was a tie on a hole (i.e., don't say "There is a 2-way tie"), you can do so with this simplified version of the formula:

```
=IF(MIN(C2:F2)=SMALL(C2:F2,2),"",
INDEX(GolferNames,MATCH(MIN(C2:F2),
C2:F2,0)))
```

Determining Winners, by Category

Suppose that your company sponsors a benefit car show and you are charged with keeping track of scores and coming up with the winners. Each column in the worksheet represents a different category of car, and each row represents a different entrant in the contest. Each cell in the table contains a score for that contestant in the appropriate categories. Your job, after tracking the scores, is to calculate the top winners in each category: first, second, and third place.

If each category will contain a unique score for each person (there are no ties), then calculating the top three scores in each category is relatively easy. Let's assume that the first three rows of the worksheet are used to show the top three winners in each category. Cell A1 contains 1 (for first place), cell A2 contains 2 (for second place), and cell A3 contains 3 (for you know which place).

The actual scoring table begins in cell A5, with column labels. Cell A5 contains the word "Names," and then cells B5:AA5 have the names of each car category. Cells A6:A100 contain the names of each contestant, and B6:AA100 contains the scores for those contestants, by category.

Enter the following formula into cell B1:

```
=INDEX($A$6:$A$100,MATCH(LARGE(B$6:B$100,$A1),B$6:B$100,0))
```

Copy the formula to the rest of the results range, B1:AA3. The formula looks at the ranking in column A (1 through 3) and then uses that to pick the first, second, and third largest values in each column. Rather than return the value, however, the value is used to pick the name of the person with that value; it is this name that is returned.

This approach, as mentioned, assumes there are no ties in the scoring table. If it is possible to have ties, then the scoring becomes much more complex and, perhaps, the best solution is to create a user-defined function in a macro. (The reason that ties make it more difficult is that the judges need to come up with a set of rules by which to break ties. These rules can vary, which means that the formulas—and user-defined functions—can vary.)

Another suggestion is to modify the way in which your scoring table is maintained. Instead of creating a large matrix (26 columns and however many contestants there are), create a small database that only has three columns: name, category, and score. You would then enter the data for each person into the database, and sort the database to get the desired winners. Simply sort first by category and then by score, and you can easily see who the top three contestants are in each category.

Determining "Highest Since" or "Lowest Since"

Alex routinely analyzes the latest building industry data, and needs to write articles about the data. Frequently he needs to highlight some new piece of data, such as "industrial building construction was the lowest since August 2007." Alex wondered if there was a way to automate this type of highlighting; if column A contains the month and year and column B contains the values for those periods, Alex would like a formula in column C that indicates "this value is the highest since April 2007" or "this value is the lowest since November 2004."

Assuming that the month and year listed in column A is really an Excel date value (and not text), you can easily create a formula to return the desired information. If you have row 1 occupied with headings for your columns, enter the following in cell C2:

```
=IF(ROW(B2)=2,"",IF(B2>MAX($B$1:B1), "this value is
the highest since " & TEXT(INDEX($A$1:A1,MATCH(MAX(
$B$1:B1),$B$1:B1,0)), "mmmm yyyy"), IF(B2<MIN($B$1:B1),
"this value is the lowest since " & TEXT(INDEX($A$1:A1,
MATCH(MIN($B$1:B1),$B$1:B1,0)), "mmmm yyyy"),"")))
```

Remember that this is a single formula, and should be entered all on one line. You can copy the formula down as many rows as necessary in column C, and it should provide the desired information. It only makes a notation in column C if the value in column B is greater than the maximum or less than the minimum of all the foregoing values in column B.

If you have quite a bit of data in your worksheet, you could notice that the formula results in long recalculation times. If this is the case, then you may want to consider using macro that will do the desired analysis and provide the appropriate information. The following macro provides looks backward through the information in column B and provides both a "lowest since" and "highest since" result in columns C and D.

```
Sub FindHiLow()
    Dim orig_cell As Range
    Dim orig_val As Integer
    Dim orig_row As Integer
    Dim rownum As Integer
    Dim newcell As Range
    Dim new_val As Integer
    Dim lowrow As Integer
    Dim hirow As Integer

    Set orig_cell = ActiveCell
    orig_row = ActiveCell.Row
    orig_val = orig_cell.Value

' find lowest
    lowrow = 0
    For rownum = orig_cell.Row - 1 To 1 Step -1
        Set newcell = Cells(rownum, 2)
        new_val = newcell.Value
        If orig_val >= new_val Then
            lowrow = rownum
            Exit For
        End If
    Next
    If lowrow = 0 Then lowrow = 1
    Cells(orig_row, 3).Value = "Lowest since " & Cells(lowrow, 1)

' find highest
    hirow = 0
    For rownum = orig_cell.Row - 1 To 1 Step -1
        Set newcell = Cells(rownum, 2)
        new_val = newcell.Value
        If orig_val <= new_val Then
            hirow = rownum
            Exit For
        End If
    Next
    If hirow = 0 Then hirow = 1
    Cells(orig_row, 4).Value = "Highest since " & Cells(hirow, 1)
End Sub
```

Finding the Dates for Minimums and Maximums

Michael has some data in two columns of a worksheet. Column A is a series of dates and column B contains expenditures for each of those dates. He needs a way to determine the date on which the minimum expenditure occurred and a way to find the date on which the maximum expenditure occurred.

At first glance you might be tempted to think you could use VLOOKUP to figure out the desired date. Unfortunately, that won't work because VLOOKUP keys off of whatever data is in the first column of your data table. Since you want to key off of the amount (looking for the minimum and maximum), you would need to switch the position of the columns, so that column A contained the amount and column B contained the dates. If you do that, you could then use the following formulas to find the dates for the minimum and maximum expenditures, respectively:

```
=VLOOKUP(MIN(A:A),A:B,2,0)
=VLOOKUP(MAX(A:A),A:B,2,0)
```

If you cannot change the data columns, then you'll need to forego VLOOKUP and use a different approach, instead. The following two formulas will work with the data as originally specified by Michael:

```
=INDEX(A:A,MATCH(MIN(B:B),B:B,0))
=INDEX(A:A,MATCH(MAX(B:B),B:B,0))
```

It is important to note that if there are duplicates of the maximum and/or minimum expenditures, these formulas will only find the dates of the first occurrences. If this is not expected to happen often, a formula such as this could be used to alert the user to the existence of duplicates:

```
=IF(COUNTIF(B:B, MAX(B:B))>1, "Multiple Maximums ",
INDEX(A:A, MATCH(MAX(B:B), B:B, 0)))
```

To make this formula work for minimums, just replace the two instances of MAX with MIN and change the word "Maximums" to "Minimums."

Returning Blanks or Asterisks From a Lookup

Bob asked if there was a way to use VLOOKUP to return blanks or asterisks if the function cannot make a match in a lookup table.

Yes, this can be done, but not without making your formula just a bit more complex. The trick is to remember that VLOOKUP can operate in one of two ways. By default, it returns the next lower value to the one being looked for, if the data table is in ascending order and if there isn't an exact match. However, you can force VLOOKUP to only return exact matches, if desired. Consider the following example:

```
=VLOOKUP(5,A1:B10,2,FALSE)
```

This example searches through the lookup table (A1:B10) looking for the value 5 in the first column of the table. If it is found, then the corresponding value from the second column is returned. If it is not found, then VLOOKUP returns an #N/A error, indicating it could not locate the value. (The FALSE value as the fourth parameter indicates you don't want approximate matches.)

The key, then, is to play off this #N/A value and build what you want returned if there isn't a match. The following formula will return a series of five asterisks if there wasn't a match in the lookup:

```
=IF(ISNA(VLOOKUP(5,A1:B10,2,FALSE)),"*****",VLOOKUP(5,A1:B10,2,FALSE))
```

The ISNA function is used to test if the result of VLOOKUP is the #N/A error. If it is, then the asterisks are returned; if not, then the lookup value is returned. If you want the formula to return "nothing," then you can use this variation:

```
=IF(ISNA(VLOOKUP(5,A1:B10,2,FALSE)),"",VLOOKUP(5,A1:B10,2,FALSE))
```

This version returns a blank string if there was not a match in the lookup table. For some uses, this may not be exactly what you want. You may find it more appropriate to return a zero, and then hide zeroes in the worksheet. (How you suppress the display of zero values in a worksheet is discussed in other *ExcelTips*.) If you'd like a zero returned, then it takes only one change:

```
=IF(ISNA(VLOOKUP(5,A1:B10,2,FALSE)),0,VLOOKUP(5,A1:B10,2,FALSE))
```

Returning Blanks with VLOOKUP

When you use VLOOKUP to return a value from a data table, the function does not differentiate between blanks and zero values in what it returns. If the source value is zero, then VLOOKUP returns 0. Likewise, if the source is blank, then VLOOKUP still returns the value 0. For some purposes, this may not do—you need to know whether the cell being looked up is blank or if it really contains a 0.

There are many different solutions that could be pursued. One solution relies on the fact that even though VLOOKUP returns a 0, it will correctly report the length of the source cell. Thus, if you use the LEN function on what is returned, if the source cell is empty the LEN function returns 0, but if the source contains a 0 then LEN returns 1 (the 0 value is 1 character in length). This means that you could use the following formula in place of a standard VLOOKUP:

```
=IF(LEN(VLOOKUP(B1,D:E,2,0))=0,"",VLOOKUP(B1,D:E,2,0))
```

In this case if the length of what VLOOKUP returns is 0, then Excel doesn't actually do a lookup—it forces a blank to be returned. Only if the length is not 0 is the actual VLOOKUP performed.

There are other variations on this same concept, each testing a different characteristic of the data being referenced and then making the decision as to whether to actually look up that data. This variation, for example, directly tests to see if the source is blank:

```
=IF(VLOOKUP(B1,D:E,2)="","",VLOOKUP(B1,D:E,2))
```

The formula can also be modified to check the source cell for multiple conditions. For instance, this variation returns a blank if the source is blank or if the source contains an #N/A error:

```
=IF(ISNA(VLOOKUP(B1,D:E,2,0))+(VLOOKUP(B1,D:E,2,0)=""),
"",VLOOKUP(B1,D:E,2,0))
```

Pulling a Phone Number with a Known First and Last Name

Kimm has a worksheet that has three columns in it. Column A contains the last name of a person, column B contains the first name, and column C contains the phone number for the individual. If Kimm knows the person's first and last name (say they are in cells F1 and F2, respectively), she wonders what sort of lookup formula she would use to return the phone number for the first person matching that first and last name.

There are actually several different formulas you could use to figure out the phone number. Most of the approaches involve using array formulas, which are always entered in a cell by using **Ctrl+Shift+Enter**. To make the formulas easier to understand, it is best to work with named ranges. For instance, set up the following names:

- Select all the last names, in column A, and give them a name such as LNames.
- Select all the first names, in column B, and give them a name such as FNames.
- Select all the phone numbers, in column C, and give them a name such as Phones.

Here are a variety of array formulas you could use to find the phone number:

```
=SUMPRODUCT(--(LNames=F2),--(FNames=F1),Phones)
=INDEX(Phones,MATCH(F1&F2,FNames&LNames,))
=INDEX(Phones,INDEX(MATCH(F1&F2,FNames&LNames,0),))
=OFFSET(C1,MATCH(F1&F2,FNames&LNames,),)
```

In addition, you can construct an array formula that relies upon the ROW function, as shown here:

```
=INDEX(Phones,SUMPRODUCT((F1&F2=FNames&LNames)*(ROW(FNames)-1)))
=INDEX(Phones,MIN(IF((FNames=F1)*(LNames=F2),(ROW(Phones)-1))))
```

If you use this approach (relying upon the ROW function), you may need to adjust the "-1" portion of the formulas to reflect the number of rows that appear before your actual data. In this instance, the row is decremented by one because the first row of the data table consists of the headers for each column; the data itself begins in row 2.

Note that all of the formulas used thus far rely upon combining the first name with the last name in order to make a comparison. This has the chance of running into "false positives" in some instances. For example, let's say that two of the names you have in your data are Thom Astonfield and Thomas Tonfield. Since the case of the letters in the names doesn't matter in these formulas, when you combine first and last names of these individuals, they are exactly the same. Thus, if you are looking for the phone number for Thomas Tonfield and his name appears in the list after Thom Astonfield, then you'll always get Thom's phone number instead of Thomas'.

To get around this potential problem, you may want to include some sort of separator between the first and last name. Using one of the formulas from earlier in the tip, all you would need to do is make an adjustment, as shown here:

```
=INDEX(Phones,SUMPRODUCT((F1&":"&F2=FNames&":"&LNames)*(ROW(FNames)-1)))
```

The addition of the colon between the first and last names acts as a separator, eliminating the likelihood of false positives.

You should also realize that if cells F1 and F2 contain a name that doesn't appear in the data at all, the information you get back from the formula will be erroneous. Depending on the variation of the function used, you may get back an actual error condition (such as #N/A or #REF) or you may get back plain wrong data. In those instances where an error condition is returned, you may want to adjust your formula to account for the possibility of not finding a match, in this manner:

```
=IF(ISERROR(INDEX(Phones,MATCH(F1&":"&F2,FNames&":"&LNames,0))),
"no phone",INDEX(Phones,MATCH(F1&":"&F2,FNames&":"&LNames,0)))
```

There are some things that you should keep in mind when using these types of formulas. First and foremost, the reliability of the information you get back is going to depend largely on the quality of the information in your data list. If your data has misspellings in it, contains blanks, is sorted in some strange order, or there are multiple entries for the same person, it can affect what the formula returns.

If you aren't too sure about the quality of your data, you may want to simply use Excel's filtering capabilities rather than a formula. Apply an AutoFilter, and you can use the first two columns of your data to pick first and last name. This will then return any phone numbers for the individual you select. It is very simple to do and makes selecting the data you need easy.

Finally, you should realize that there are other approaches you can use to addressing the problem. For instance, you could create a formula that uses the DGET function, but doing so would require the addition of a small criteria table to your worksheet or workbook. Since Kimm specified that she couldn't add intermediate results in a worksheet, then an editorial decision was made to not include the DGET function as a solution since it would require the addition of the criteria table.

In addition, if you are comfortable with using macros, you could also create a user-defined function that would examine the data and return the requested phone number. The benefit to using such an approach is that it provides you greater flexibility in processing the information that the function actually returns.

Looking Up Names when Key Values are Identical

Jim described a situation where he has a list of employee names and their salaries. He wants to determine who the five highest-paid employees are. He uses the LARGE function to identify the five largest salaries, and then tries to use VLOOKUP to return the names belonging to those salaries. This works fine unless there are duplicates in the top five salaries (people get paid the same salary). If there are, then VLOOKUP only returns the name of the first employee at that salary.

To return all the proper names, there are a couple things you could do. One method would be to bypass using a formula altogether. Instead, you could use the AutoFilter feature in Excel:

1. Select any cell in your data table.
2. Choose Data | Filter | AutoFilter or, if you are using Excel 2007 or Excel 2010, display the Data tab of the ribbon and click the Filter tool. Excel adds drop-down arrows at the right of each column header in the table.
3. Use the drop-down list at the top of the salaries column to choose Top 10. (In Excel 2007 and Excel 2010, use the drop-down list to choose Number Filters and then Top 10.) Excel displays the Top 10 AutoFilter dialog box.

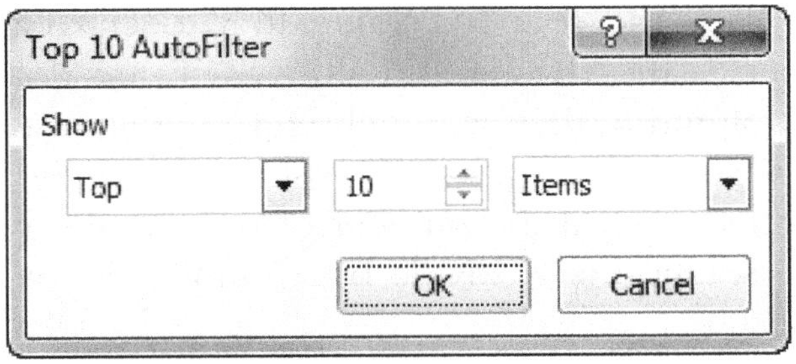

The Top 10 AutoFilter dialog box.

4. Adjust the center control from 10 to 5.
5. Click on OK. Excel displays the top five salaries in the list.

When you follow these steps, you may actually end up with more than five records visible, particularly if there are ties in the employee salaries. The filter identifies the top five salaries and then displays all the records with salaries matching those.

If you don't want to use the AutoFilter, another option is to simply make sure that there is something unique about each of the records in your employee list. For instance, if the employee names are in column B and the salaries are in column C, then you could use the following formula in column A to make each record unique:

```
=C2+ROW()/100000000
```

This will add the row number divided by 100,000,000 and will make a unique value. If you have (for example) identical salaries of 98,765.43 in rows 2 and 49 in column A they will be:

```
98765.43000002
98765.43000049
```

The large number (100,000,000) is so that if you had an identical number in row 65536, you would get:

```
98765.43065536
```

And even in this case the rounded value to 2 decimal places would still be the real number. If the LARGE and VLOOKUP are done with the "non-unique" values in column A, then you will return the largest salaries (and their associated people), based on the person's position within the list. The only time this approach won't work is if the number of rows in the worksheet exceeds 999,999 rows. The solution is to divide by a number larger than 100,000,000 (such as 1,000,000,000).

A third approach is to use the RANK and COUNTIF functions to return a unique "ranking" for each value in the list of salaries. If the salaries are in the range B1:B50, enter the following in cell C1 and copy it down the range:

```
=RANK(B1,$B$1:$B$50)+COUNTIF($B$1:B1,B1)-1
```

You can now use INDEX on the ranking values to return the name associated with each salary.

Finally, a fourth approach is to create a macro that can return the desired information. There are many ways that a macro could be implemented; the following is just one of them:

```
Function VLIndex(vValue, rngAll As Range, _
  iCol As Integer, lIndex As Long)
    Dim x As Long
    Dim lCount As Long
    Dim vArray() As Variant
    Dim rng As Range
    On Error GoTo errhandler

    Set rng = Intersect(rngAll, rngAll.Columns(1))
    ReDim vArray(1 To rng.Rows.Count)
```

```
        lCount = 0
        For x = 1 To rng.Rows.Count
            If rng.Cells(x).Value = vValue Then
                lCount = lCount + 1
                vArray(lCount) = rng.Cells(x).Offset(0, iCol).Value
            End If
        Next x

        ReDim Preserve vArray(1 To lCount)
        If lCount = 0 Then
            VLIndex = CVErr(xlErrNA)
        ElseIf lIndex > lCount Then
            VLIndex = CVErr(xlErrNum)
        Else
            VLIndex = vArray(lIndex)
        End If
    errhandler:
        If Err.Number <> 0 Then VLIndex = CVErr(xlErrValue)
    End Function
```

The parameters passed to this user-defined function are the value, the range of cells to lookup in, the "offset" from this range for the lookup (the number of columns to the right is positive, to the left is negative) and the number of the duplicate (1 is first value, 2 the second, and so on).

To use it, for example's sake, assume A1:B1 contain column headers, A2:A100 contains the salaries, and B2:B100 contains the employee names. In cell E2 you can enter the following to determine the largest salary in the table:

```
=LARGE($A$2:$A$100,ROW()-1)
```

In cell F2 you can enter the following formula to determine if the row has any duplicates and keep track of the current "value" of that duplicate:

```
=IF(E2=E1,1+F1,1)
```

In cell G2 you can use the following formula, which invokes the user-defined function:

```
=VLIndex(E2,$A$2:$A$100,1,F2)
```

Copy cells E2:G2 to E3:G6, and you will have (in column G) the names of the employees with the five largest salaries.

Displaying Letter Grades

Some teachers use Excel worksheets to calculate grades for students. Doing so is quite easy, as you sum the results of various student benchmarks (quizzes, tests, assignments, etc.) and then apply whatever calculation is necessary to arrive at a final numeric grade.

If you do this, you may wonder how you can convert the numeric grade to a letter grade. For instance, you may have a grading scale defined where anything below 52 is an F, 52 to 63 is a D, 64 to 74 is a C, 75 to 84 is a B, and 85 to 99 is an A.

There are several ways that a problem such as this can be approached. First of all, you could use nested IF functions within a cell. For example, let's assume that a student's numeric grade is in cell G3. You could use the following formula to convert to a letter grade based on the scale shown above:

```
=IF(G3<52,"F",IF(G3<64,"D",IF(G3<75,"C",IF(G3<85,"B","A"))))
```

While such an approach will work just fine, using nested IF functions results in the need to change quite a few formulas if you change your grading scale. A different approach that is also more flexible involves defining a grading table and then using one of the LOOKUP functions (LOOKUP, HLOOKUP, and VLOOKUP) to determine the proper letter grade.

As an example, let's assume that you set up a grading table in cells M3:N7. In cell M3 you place the lowest possible score, which would be a zero. To its right, in cell N3, you place the letter grade for that score: F. In M4 you place the lowest score for the next highest grade (53) and in N4 you place the corresponding letter grade (D). When you are done putting in all five grade levels, you select the range (M3:N7) and give it a name, such as GradeTable. (How you name a range of cells is covered in other issues of ***ExcelTips***.)

Now you can use a formula such as the following to return a letter grade:

```
=VLOOKUP(M22,GradeTable,2)
```

The beauty of using one of the LOOKUP functions in this manner is that if you decide to change the grading scale, all you need to do is change the lower boundaries of each grade in the grading table. Excel takes care of the rest and recalculates all the letter grades for your students.

When you put together your grading table, it is also important that you have the grades—those in the GradeTable—go in ascending order, from lowest to highest. Failure to do so will result in the wrong formula results using VLOOKUP.

Determining a Name for a Week Number

Theo uses an Excel worksheet to keep track of reservations in his company. The data consists of only three columns. The first is a person's name, the second the first week number (1-52) of the reservation, and the third the last week number of the reservation. People can be reserved for multiple weeks (i.e., start week is 15 and end week is 19). Theo needs a way to enter a week number and then have a formula determine what name (column A) is associated with that week number. The data is not sorted in any particular order, and the company won't let Theo use a macro to get the result (it has to be a formula).

Theo's situation sounds simple enough, but it is filled with pitfalls when devising a solution. Looking at the potential data (as shown in the following figure) quickly illustrates why this is the case.

	A	B	C
1	**Name**	**Start**	**End**
2	Bill	2	3
3	Fred	11	13
4	George	9	9
5	Carl	10	11
6	Donna	7	8
7	Jerry	4	4
8	Hank	15	18
9	Robert	16	16
10	Wilma	20	21
11			

Potential data for Theo's problem.

Notice that the data (as Theo said) is not in any particular order. Note, as well, that there are some weeks where there are no reservations (such as week 5 or 6), weeks where there are multiple people (such as week 11 or 16), and weeks where there is someone reserved, but the week number doesn't show up in column B or C (such as week 12 or 17).

Before starting to look at potential solutions, let's assume that the week you want to know about is cell E1. You should name this range as Query. Further, name the range that contains people's names (in this example, cells A2:A10) as ResNames, the starting weeks (B2:B10) as StartWeeks, and the ending weeks (C2:C10) as EndWeeks. Finally, define a name for the entire table (A2:C10), such as MyData. This naming, while not strictly necessary, will make understanding the formulas much easier.

One potential solution is to add what is commonly referred to as a "helper column." Add the following to cell D2:

```
=IF(AND(Query>=B2,Query<=C2),"RESERVED","")
```

Copy the formula down, for as many cells as there are names in the table. (For example, copy it down through cell D10.) When you place a week number in cell E1, then the word "RESERVED" appears to the right of any reservation that involves that week number. It is also easy to see if there are multiple people reserved for that week or if there are no people reserved for that week. You could even apply an AutoFilter and select to only show those records with the word "RESERVED" in column D.

You can, if desired, forego the helper column and consider using conditional formatting to display who is reserved for a desired week. Simply select the names in column A and add a conditional formatting rule that uses the following formula:

```
=AND(Query>=B2,Query<=C2)
```

(How you enter conditional formatting rules has been described extensively in other issues of ***ExcelTips.***) Set the rule so that it changes the shading (pattern) applied to the cell, and you'll easily be able to see which reservations apply to the week you are interested in.

Another approach is to use an array formula. Select a few more cells than the number of overlapping reservations you expect, and then enter the following into those cells by pressing **Ctrl+Shift+Enter**:

```
=IFERROR(INDEX(ResNames,LARGE((StartWeeks<=Query)*(EndWeeks>=Query)*(ROW(Res
Names)),ROW()-1)-1),"")
```

This approach will work only if you are using Excel 2007 or Excel 2010. The reason is simple: The IFERROR function is not available in earlier versions of Excel.

When picking the number of cells you want this array formula to occupy, look at, for instance, the number of people that may be reserved over week 11. In the example shown in this tip, it is 2 people. Select more than that number of cells and then put the array formula in those cells. If you expect that you might have 20 people potentially booked for the same week, then you'll want to pick a larger number of cells, such as 20 or 30. Just select the cells, put the formula in the Formula bar, and then press **Ctrl+Shift+Enter**.

Finally, you really should consider revising how your data is laid out. You could create a worksheet that has week numbers in column A (1 through 52 or 53) and then place names in column B. If a person was reserved for two weeks, their name would appear in column B twice, once beside each of the two weeks that they reserved.

With your data in this format, you could easily scan the data to see which weeks are available, which are taken, and who they are taken by. If you want to do some sort of lookup, it is easy to use the VLOOKUP function based on the week number, since it is the first column of the data, in sorted order.

Complex Lookup Formulas

Eddie has a series of part numbers in the format 123/45678 or 011/00345. He needs to look up the name of the corresponding part number from a different table. This table consists of three columns. The first column contains the portion of the part number before the slash (such as 123 or 011) and the second column contains the portion after the slash (such as 45678 or 00345). The third column contains the name that Eddie want to look up.

Part of the problem Eddie is having is that in the table, the two columns that each contains a portion of the part number are numeric values. Thus the full part number isn't 011 and 00345, but 11 and 345. Eddie is wondering how to put together a lookup formula for the part numbers (123/45678 or 011/00345) to return the proper part name from the table.

To start with, you need to make some changes to the table that contains the part numbers. The first two columns should be formatted as text, not as numbers. This will treat the values in those columns as text, so that when you enter "011" in the first column, it retains the leading zero. If you have already entered the value as 011 before formatting the column as text, it will still be displayed as 11 (without the leading zero). You will need to not only change the format of the column, but also re-enter any part-number fragments that contain leading zeros.

Next you need to make sure that your part number table includes column headers. For this example, make sure the three column headers are Left, Right, and Name. (You can make these headers bold and underlined, which helps set them off from the values in each column.) Then, create names for the individual columns by following these steps if you are using Excel 2007 or Excel 2010:

1. Select the cells in the part number table. Make sure you also select the newly created headers for each column of the table.

17	**Left**	**Right**	**Name**
18	008	85343	Chair
19	011	34971	Table
20	011	25876	Desk
21	123	45678	Vase
22	233	76398	Lamp
23	548	65873	Bed

The selected cells of the part-number table.

2. Display the Formulas tab of the ribbon.
3. Click the Create from Selection tool, in the Defined Names group. Excel displays the Create Names from Selection dialog box.

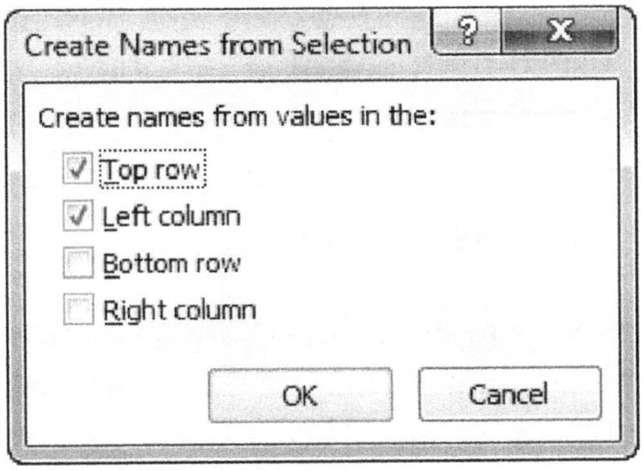

The Create Names from Selection dialog box.

4. Make sure that only the Top Row check box is selected.
5. Click OK.

If you are using an older version of Excel, follow these steps instead:

1. Select the cells in the part number table. Make sure you also select the newly created headers for each column of the table.
2. Choose Insert | Name | Create. Excel displays the Create Names dialog box.

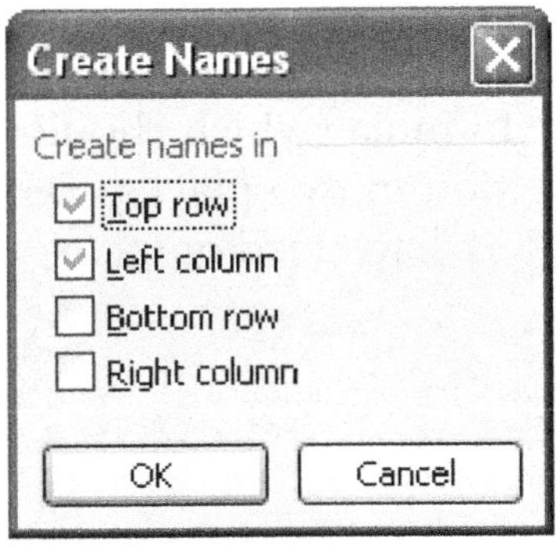

The Create Names dialog box.

3. Make sure that only the Top Row check box is selected.
4. Click OK.

With your part-number table prepared, you are now ready to look up part numbers. In the cells just to the right of your lookup values (123/45678 and 011/00345) you are going to enter an array formula. Assuming that the first part number is in cell A1, you would enter the following array formula in cell B1:

```
=INDEX(Name,MATCH(A1,Left&"/"&Right,0))
```

Remember that to denote this as an array formula you enter it by pressing **Shift+Ctrl+Enter**. The formula then appears in the Formula Bar with {braces} around it. You can copy the formula down to the other cells in column B, as necessary.

The formula works by pulling the value from the Name column of the table when the corresponding Left-column value concatenated with a slash and the Right-column value matches whatever is in cell A1. If there is no match, the formula returns a #N/A error value, otherwise it returns the desired part number.

Note that this approach will work only if you format the first two columns of your part-number table as text and make sure that the various cells contain any leading zeros. If, for some reason, you cannot format the first two columns of the table in this way (perhaps there are too many of them), then you need to change the lookup formula you are using:

```
=INDEX(Name,MATCH(A1,TEXT(Left,"000")&"/"&TEXT(Right,"00000"),0))
```

Note that this version of the formula (which should still be entered as an array formula) uses the TEXT function in two places, to convert the values from the Left and Right columns so they have leading zeros.

5

Using Lookup Functions with Dates

Macro for Month Name

Brian is looking for a macro that returns the full name of the current month, such as July, August, etc. Before getting to the macros, it should be mentioned that depending on your needs, you could get the desired information with one of several formulas. Perhaps the easiest formula is the following:

```
=Today()
```

Place the formula into a cell, and you end up with today's date. Format the cell using a custom format, and you end up with the full month name. The custom format is applied by using these steps:

1. Select the cell containing the formula.
2. Display the Format Cells dialog box. (In Excel 2007 or Excel 2010 display the Home tab of the ribbon and click the small icon at the lower-right corner of the Number group. In older versions of Excel choose Cells from the Format menu.)
3. Make sure the Number tab is displayed.
4. In the list of format categories, select Custom.

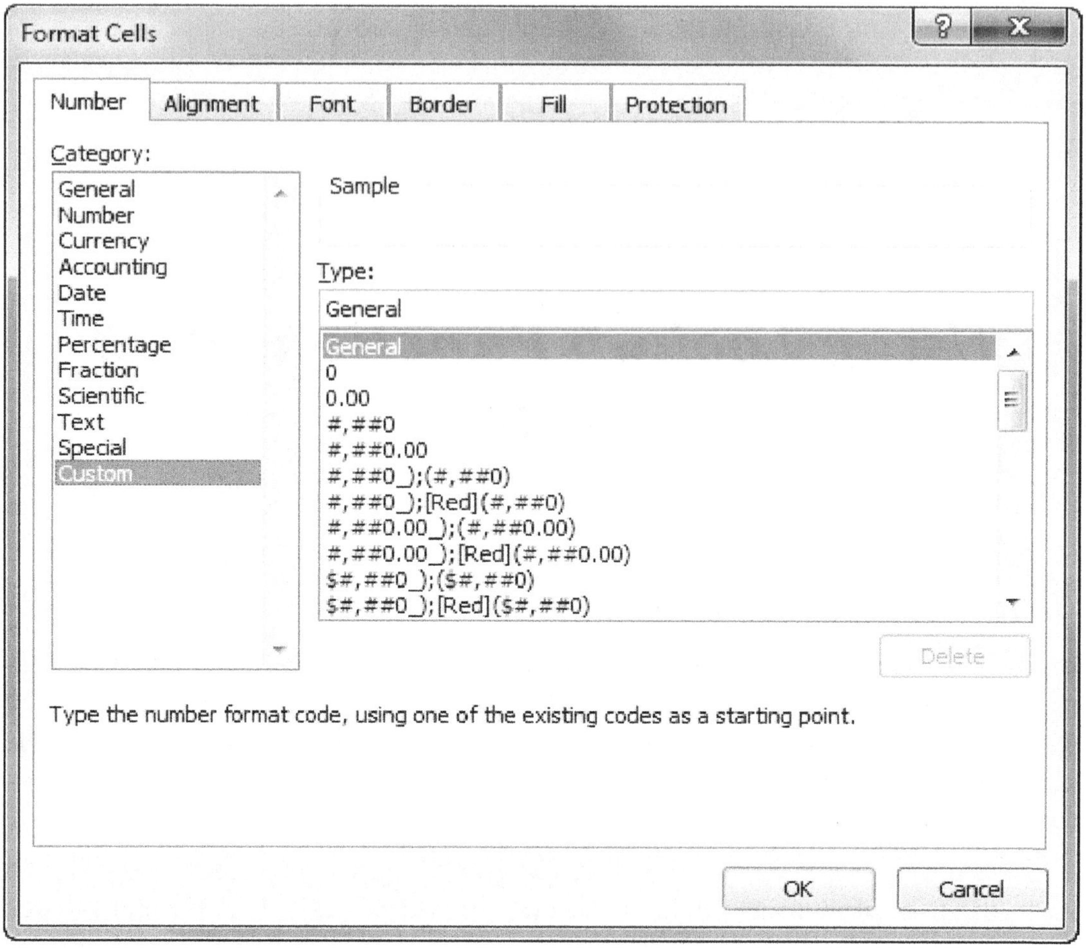

The Number tab of the Format Cells dialog box.

5. In the Type box, enter **mmmm**.
6. Click OK.

Another formulaic approach is to use the following in a cell:

```
=Text(Today(),"mmmm")
```

No special formatting is required; the formula returns the text of the full month name for whatever today is. Finally, you could use an even longer formula that simply picks the month name from a list of months:

```
=CHOOSE(MONTH(NOW()),"January","February",
"March","April","May","June","July",
"August","September","October","November",
"December")
```

Remember that this is a single formula; it goes all in one cell.

Which brings us, finally, to the macros. If you want a macro that returns the month name in the current cell, you are looking for a user-defined function:

```
Function MonthName()
    Application.Volatile
    MonthName = Format(Date, "mmmm")
End Function
```

This simple two-line macro dynamically returns the month name for whatever the current date is. Just put this formula in a cell:

```
=MonthName()
```

Remember—since you've just added a macro to your workbook, you'll be asked whenever you open your workbook if you want to enable macros. If you don't want to see this question all the time, you should use one of the formulaic approaches presented earlier.

Automatically Advancing by a Month

Jim has a need to advance the date in a particular cell by one month at midnight on the 14/15 of each month (00:00:00 on the 15th), and he wondered how it could be done.

As with many problems in Excel, the answer depends on the nature of the data involved and exactly what you want to do. If the date in the cell is today's date, and you simply want to have the cell display the current month up through the 14th, and then next month after that, then you can use a formula such as the following:

```
=CHOOSE(IF(DAY(NOW())>14,MONTH(NOW())+1,MONTH(NOW())),
"January","February","March","April","May","June",
"July","August","September","October","November",
"December","January")
```

This formula returns the name of a month, not a date. If you prefer to have a date returned, you can use this formula:

```
=IF(DAY(NOW())>14,DATEVALUE(IF(MONTH(NOW())=12,1,
MONTH(NOW())+1) & "/" & DAY(NOW()) & "/" & IF(MONTH(
NOW())=12,YEAR(NOW())+1,YEAR(NOW())))),NOW())
```

Both of these formulas account for the "end of year wrap-around" when you advance from December to January. A shorter version of this last formula can be created if you use the DATE function instead of the DATEVALUE function:

```
=DATE(YEAR(NOW()),MONTH(NOW())+((DAY(NOW())>14)*1),1)
```

This formula, unlike the DATEVALUE example, always returns a date that is the first day of any given month.

If you really want to advance the value of a particular date in a cell, then you must use a macro to do the task. Further, you must make sure that the macro only runs once a month, at a particular time on a particular day. For instance, if you wanted the macro to run at 00:00:00 on the 15th of each month, you would need to set up the macro so that it checked the date and time, and then ran at that particular date and time. You would also need to make sure that the workbook containing the macro was open over that date and time.

The following macro will fetch the date from a cell and increase it by a month. The macro assumes that you have a named range, DateCell, which refers to the cell to be updated.

```
Sub IncreaseMonth()
    Dim dDate As Date
    dDate = Range("DateCell").Value
    Range("DateCell").Value = _
      DateSerial(Year(dDate), _
      Month(dDate) + 1, Day(dDate))
End Sub
```

To make sure that the macro runs at the appropriate time, you would need another macro. The following macro is designed to be run whenever the workbook is opened:

```
Private Sub Workbook_Open()
If Day(Now) = 14 Then
    Application.OnTime ("23:59:59"), "IncreaseMonth"
End If
End Sub
```

Notice that this particular macro sets the OnTime method so that it runs the IncreaseMonth macro at 23:59:59 on the 14th. This date and time was chosen because it is easier to catch than is 00:00:00 on the 15th.

Remember that the IncreaseMonth macro will only run if you open the workbook on the 14th, and then leave the workbook open until the 15th.

Adding Ordinal Notation to Dates

When developing a workbook, you may have a need to place suffixes such as "st, nd, rd, or th" at the end of dates, as in "9th March." Unfortunately, there is no way to do this using the built-in date formats you can apply to individual cells. You can create custom formats for each of the four suffix types, if desired, but they would have to be applied individually based on the contents of the cell itself.

The only other option is to use some sort of conversion formula. These are easy enough to put together, but the resulting cell will not contain a true Excel date, but text. This precludes the cell contents from being used in other date-related functions. The following is an example of the type of conversion formula you can use:

```
=DAY(A1)&IF(OR(DAY(A1)={1,2,3,21,22,23,31}),
CHOOSE(1*RIGHT(DAY(A1),1),"st","nd ","rd "),"th")
&TEXT(A1,"mmmm, yyyy")
```

There are others, but they all essentially do the same thing—pull the various parts of a date apart and put them back together with the proper suffix.

If you prefer, you can also create a macro function that would return a properly formatted date, with the ordinal suffix. The following is one such macro:

```
Function OrdinalDate(myDate As Date)
    Dim dDate As Integer
    Dim dText As String
    Dim mDate As Integer
    Dim mmmText As String

    dDate = Day(myDate)
    mDate = Month(myDate)

    Select Case dDate
        Case 1: dText = "st"
        Case 2: dText = "nd"
        Case 3: dText = "rd"
        Case 21: dText = "st"
        Case 22: dText = "nd"
        Case 23: dText = "rd"
        Case 31: dText = "st"
        Case Else: dText = "th"
    End Select

    Select Case mDate
        Case 1: mmmText = " January"
        Case 2: mmmText = " February"
        Case 3: mmmText = " March"
        Case 4: mmmText = " April"
        Case 5: mmmText = " May"
        Case 6: mmmText = " June"
        Case 7: mmmText = " July"
        Case 8: mmmText = " August"
        Case 9: mmmText = " September"
        Case 10: mmmText = " October"
        Case 11: mmmText = " November"
        Case 12: mmmText = " December"
    End Select

    OrdinalDate = dDate & dText & mmmText
End Function
```

You use the macro by simply invoking it within a cell formula. For example, if you have a date stored in cell B7, you can use the following in any other cell:

```
=OrdinalDate(B7)
```

Date for Next Wednesday

Sam has a date in cell A1 and he would like to calculate the date of the following Wednesday in cell B1. He wonders what formula he should use for the calculation.

There are actually many formulas you can use, and the one you pick is pretty much up to you. Here is a good representative sampling of formulas you can use:

```
=IF(WEEKDAY(A1)<=4,A1+4-WEEKDAY(A1),A1+11-WEEKDAY(A1))
=A1+WEEKDAY(A1, 1)+CHOOSE(WEEKDAY(A1, 1), 2, 0, -2, -4, 1, -1, -3)
=A1+CHOOSE(WEEKDAY(A1),3,2,1,0,6,5,4)
=A1-MOD(WEEKDAY(A1)-5,7)+6
=A1+MOD(4-WEEKDAY(A1),7)
```

These formulas return a date that represents next Wednesday, provided that the date in cell A1 isn't a Wednesday to begin with. For instance, if the date in A1 is 5/26/10 (a Wednesday), then each of these will return 5/26/10. However, if the date is 5/27/10 then the formula returns 6/2/10.

If you want a formula that will return the next Wednesday even when the starting date is a Wednesday, then you should rely on a different formula. Choose from one of these:

```
=A1+IF(WEEKDAY(A1,1)=4,7,IF(WEEKDAY(A1,1)<4,4-WEEKDAY(A1,1),11-
WEEKDAY(A1,1)))
=IF(WEEKDAY(A1)<4,A1+4-WEEKDAY(A1),A1+11-WEEKDAY(A1))
=IF(WEEKDAY(A1)=4,A1+7,A1+MOD(4-WEEKDAY(A1),7))
=A1+MOD(4-WEEKDAY(A1),7)+7*(0=MOD(4-WEEKDAY(A1),7))
=A1+7-MOD(4+WEEKDAY(A1,2),7)
=A1+4-WEEKDAY(A1)+IF(WEEKDAY(A1)<4,0,7)
=A1+CHOOSE(WEEKDAY(A1),3,2,1,7,6,5,4)
=A1+(7-MOD(WEEKDAY(A1,3)-2,7))
=A1+4-WEEKDAY(A1)+7*(WEEKDAY(A1)>=4)
=A1-MOD(WEEKDAY(A1)-4,7)+7
=A1+4+((WEEKDAY(A1)>=4)*7)-WEEKDAY(A1)
=A1+MOD(10-WEEKDAY(A1),7)+1
=A1+IF(WEEKDAY(A1) < 4,4,11)-WEEKDAY(A1)
=CEILING(A1-4,7)+4
=A1+6-MOD(A1+2,7)
```

It should be noted that the last two of these formulas works just fine on the PC but won't work as expected on the Mac. This is because the basis date used for date serial numbers is different on the Mac than it is on the PC, and thus the computations—which operate on the underlying serial numbers for the dates—return different values on each platform.

Calculating Future Workdays

Using Excel to calculate a date in the future is rather easy. If you have a cell (such as C3) that contains a starting date, you can simply use a formula such as the following in a different cell:

```
=C3 + 3
```

If you format the cell with the formula as a date, it will be three days in the future.

When you want to calculate workdays, the task gets trickier. For instance, you only want to return a date that is between Monday and Friday. If the starting date was a Thursday, this means the return date should be Monday, even though Sunday is the real day that is three days hence.

One quick way to figure a date three workdays in the future is to use the CHOOSE worksheet function. For instance, let's say you have the issue date for a document, and you store that date in cell B5. If you want cell B6 to show a date three workdays later, then you would place the following formula in cell B6 and make sure it is formatted as a date:

```
=B5 + CHOOSE(WEEKDAY(B5), 3, 3, 3, 5, 5, 5, 4)
```

This formula assumes that workdays are Monday through Friday. You can tinker with it to pick a different five-day workweek, if desired.

If you also want your formula to take holidays into account, then you must get a bit more creative. For these instances you can use the WORKDAY function. This function is built into Excel 2007 and Excel 2010, but in previous versions requires the Analysis ToolPak add-in. This means that you must make sure the Analysis ToolPak add-in is loaded before you can use WORKDAY. You can check if it is loaded by choosing Add-Ins from the Tools menu. Once the add-in is loaded, you could use the following formula in cell B6 to calculate the target date:

```
=WORKDAY(B5,3)
```

After you format the cell as a date, it will show the date three workdays in the future. To include holidays, the simplest way is to set up your holidays in the worksheet. For instance, you might put your company holidays in the worksheet in cells K4 through K10. Then, select the cells and give them a name, such as Holidays. You can now use your holiday range in the WORKDAY function. Change the formula in cell B6 so it looks like this:

```
=WORKDAY(B5,3,Holidays)
```

Now the function will always take your holidays into account when returning a date three workdays in the future.

Calculating the First Business Day of the Month

Dan is setting up a worksheet for use in his company and needs a way to calculate the first business day of a calendar month. He knows how to calculate the first day of a month, but that day could possibly fall on a weekend or on a holiday.

There are several ways you can calculate the first weekday of a given month. The following formula is a good place to start:

```
=DATE(YEAR(A1),MONTH(A1),1)+CHOOSE(WEEKDAY(DATE(YEAR(A1),MONTH
(A1),1)),1,0,0,0,0,0,2)
```

This formula assumes that you want the first weekday for the month of whatever date is in cell A1. Thus, if the date in cell A1 is 26 March 2011, the formula assumes you want the first weekday for March 2011.

You can make the formula much simpler by using the EOMONTH and WORKDAY functions, in this manner:

```
=WORKDAY(EOMONTH(A1,-1), 1)
```

(This formula works just fine in Excel 2007 and Excel 2010. If you are using an older version of Excel, you'll need to make sure you have the Analysis ToolPak installed on your system.)

The EOMONTH function returns the last day of the month prior to the date in cell A1, and then the WORKDAY function returns the first weekday in the following month. The function is put together like this (using EOMONTH) because WORKDAY is a bit quirky; it assumes that the first parameter is a valid work day. This means that it doesn't check whether the date used in the parameter falls in the Monday through Friday work week. By forcing the parameter to be a day before the month in which you want to determine the first work day and setting the second parameter to 1, the function returns the first actual weekday for the desired month.

The WORKDAY function also allows you to optionally specify that holidays should be taken into account. The easiest way to do this is to set up a list of holiday dates, in a cell range, and then name that range. (How you name a range is discussed in other issues of ***ExcelTips***.) If you give the range a name such as MyHolidays, then you can use the following version of the formula:

```
=WORKDAY(EOMONTH(A1,-1), 1, MyHolidays)
```

Here's a neat trick if you want to calculate the beginning work day for any given month in a year. Assume that cell A1 contains a year, such as 2011, and that cell A2 contains a month number, 1 through 12, representing the month for which you want the beginning work day. You can then use this formula to get the desired date:

```
=WORKDAY(DATE(A1,1,0), A2, MyHolidays)
```

The way the DATE function is used in this formula returns the last day of the year prior to the year specified in cell A1. Then the value in A2 is used to move forward that number of months, and MyHolidays (as before) accounts for any holidays that may occur in that month.

Determining Business Quarters from Dates

Robert is looking for a way to determine the business quarter in which a particular date falls. For example, if cell A1 has 2/15/12, he would want cell B1 to contain a formula that returned Q1-12. Similarly, if cell A1 has 8/1/11, he would want cell B1 to return Q3-11.

There are literally dozens of ways that you can determine a straight business quarter from a date. A few of these methods are worth examining, and they are easily modified to adapt to any specific needs. What is assumed here is that you really want to use a self-contained formula, rather than using a lookup table or a VBA macro. Both of those approaches will work just fine, but it is assumed that the simple formulas will work best for your purposes.

The key factor in determining the business quarter is to look at the month of the date. One way to do that examination is to use a formula that relies on the CHOOSE function. Consider the following:

```
=CHOOSE(MONTH(A1),"Q1","Q1","Q1","Q2","Q2",
"Q2","Q3","Q3","Q3","Q4","Q4","Q4") & "-"
& RIGHT(YEAR(A1),2)
```

This is a single formula; it is rather long, providing a choice for each of the 12 months in the year. Each month returns the quarter portion of the result, and then the text for the year is appended.

Another way is to rely on IF statements to determine the quarter. This is done in the following manner:

```
=IF(MONTH(A1)<=3,"Q1",IF(MONTH(A1)<=6,"Q2",
IF(MONTH(A1)<=9,"Q3","Q4")))&"-"&TEXT(A1,"yy")
```

The IF statements check the month to see its relation to the various boundaries for the quarters (3, 6, and 9) and then assigns a proper quarter (Q1, Q2, Q3, or Q4) based on the result. A dash and the last two digits of the year is then appended to the quarter.

You can make the formula even shorter by calculating the quarter directly based upon the month. For instance, the following will take the month and return a value of 1 to 4 based on the month:

```
=INT((MONTH(A1)-1)/3)+1
```

This formula can be incorporated into a larger formula in this way:

```
="Q" & INT((MONTH(A1)-1)/3)+1 & "-" & RIGHT(YEAR(A1),2)
```

Specifying Different Weekends with NETWORKDAYS

Sunil can use the NETWORKDAYS function to return the number of regular business days between two dates. The function assumes that Saturday and Sunday are not work days, but in Sunil's organization only Sunday counts as a non-work day. He wonders if there is a way to use NETWORKDAYS and specify that only Sunday should be excluded from the count returned.

You can determine this by using a formula based on the NETWORKDAYS function. Assuming that the starting date is in A1 and the ending date is in B1, the following formula examines the days between the two dates and essentially return a count of non-Sunday days in that range:

```
=NETWORKDAYS(A1,B1)+SUMPRODUCT(--(WEEKDAY
(ROW(INDIRECT(A1&":"&B1)))=7))
```

Of course, since Sundays are the only day of the week being excluded, you could simply skip the use of NETWORKDAYS and use SUMPRODUCT to figure out if the day should be counted or not:

```
=SUMPRODUCT(--(WEEKDAY(ROW(INDIRECT(A1&":"&B1)))>1))
```

If you expect that there may be holidays in the range, and that those holidays are in the named range "holidays," then you'll need to go back to using NETWORKDAYS in the formula:

```
=NETWORKDAYS(A1,B1,holidays)+SUMPRODUCT(--
(WEEKDAY(ROW(INDIRECT(A1&":"&B1)))=7),--
(NOT(ISNUMBER(MATCH(ROW(INDIRECT(A1&":"&B1))
,holidays,0)))))
```

Other variations of what constitutes the work days in a week have been covered by Chip Pearson at his site:

```
http://www.cpearson.com/excel/betternetworkdays.aspx
```

Finding the Date Associated with a Negative Value

Stuart has a series of readings in a worksheet. In the first column he has dates associated with the readings and in the second column he has the actual readings. Stuart would like to have a formula that will return the first date at which a reading became negative. In other words, the formula should look for the first value that is negative in the second column and then return the date associated with that value. There can be multiple negative values in the second column, but he needs only the date associated with the first negative value.

There are a number of ways that this problem can be approached. All of the methods presume that the dates in column A are in ascending order and that the readings in column B are not in any type of discernable order. (In other words, the readings could bounce above and below 0 on any given date.)

Provided that you have some control over the layout of the worksheet, you could add an intermediate work column in column C, used to indicate when a value is negative. Simply place a formula like this in column C, to the right of each reading:

```
=IF(B1<0,A1,"")
```

This formula returns the date in column A if the value in B is below 0 (negative), otherwise it returns nothing. All you then need to do is look for the minimum value in column C:

```
=MIN(C:C)
```

Format the result as a date, and it represents the date at which the readings first became negative.

Another approach is to forego the use of the intermediate column and use an array formula to determine the date. Assuming the data is in the range A1:B42, you can use any of the following formulas:

```
=MIN(IF(B1:B42<0,A1:A42,""))
=OFFSET($A$1,MATCH(TRUE,$B$1:$B$42<0,0)-1,,,)
=INDEX(A:A,MIN(IF(B1:B42<0,ROW(B1:B42))))
=INDEX(A1:A42,MATCH(TRUE,B1:B42<0,0))
=INDIRECT("A"&MIN(IF(B1:B42<0,ROW(B1:B42))),TRUE)
```

Remember that these are all array formulas, so you need to enter whichever one you choose by pressing **Shift+Ctrl+Enter**. Format the result as a date, and it is the answer you seek.

If you prefer, you could also use a simple macro to determine the date:

```
Function GetFirstNegative(rngdata)
    Dim c As Variant

    For Each c In rngdata
        If c < 0 Then
            GetFirstNegative = c.Offset(0, -1)
            Exit Function
        Else
            GetFirstNegative = "All Data is Positive"
        End If
    Next
End Function
```

In your worksheet, you would use this user-defined function in this manner:

```
=GetFirstNegative(B1:B42)
```

6

Looking Up Random Data

Picking Different Random Numbers from a Range

Let's say that you have a list of employees (or products, or widgets, or whatever) and that you want to pick two items from this list, at random. There are a couple of different worksheet functions you can use to pick items from the list, such as RANDOM and RANDBETWEEN.

Many people use the RANDBETWEEN function because it is very easy and provides a random number within a range. Thus, if you have 25 items in your list, RANDBETWEEN can return a number between 1 and 25, which can then correspond to items in the list.

For example, let's say that your list of items is in the range A2:A26, and that you give this range the name Items. You could then put the following formula in cell C2 to return a number between 1 and the number of Items:

```
=RANDBETWEEN(1,ROWS(Items))
```

Copy this formula to cell C3, and you now have two random numbers that represent items from the list. In cell D2 and D3 you could put formulas like this to get the actual names from the list:

```
=INDEX(Items,C2)
```

The only problem with this approach is that it is possible for both instances of RANDBETWEEN (cells C2 and C3) to return the same value, and thus you end up with the same item selected twice from your list.

One way to work around this potential problem is to actually select three items from the list instead of two. If the first two items are the same, then the third can be used as a "fallback" item to provide the unique second. The method is not foolproof, as it is possible—but quite improbable—that all three will be the same.

A different approach to selecting items from the list would be to assign each item its own random value, and then select based on the highest number in the series. The RANDOM function returns a random value between 0 and 1. In each cell of column B, just to the right of each item in column A, put this formula:

```
=RAND()
```

Select the range of cells (B2:B26) and name the range, using a name such as ItemNums. You can then determine the first random name from the list by using the following array formula:

```
=OFFSET(A$1,SUM((LARGE(ItemNums,1)=(ItemNums))*ROW(ItemNums))-1,0)
```

In order to signify that this is an array formula, enter it by pressing **Shift+Ctrl+Enter**. The formula should return a single name. You can then use the following array formula to return the second name:

```
=OFFSET(A$1,SUM((LARGE(ItemNums,2)=(ItemNums))*ROW(ItemNums))-1,0)
```

The reason that using the RAND function approach works better than using RANDBETWEEN is because the chance that RAND will return two identical values is infinitesimally small, whereas the chances of RANDBETWEEN doing so is much higher.

Selecting Random Names

A common task for many people is to pick a number of random names from a large list. For instance, you may be running a contest for your community, and you have 1,000 people that have entered. With their names in each row of a table, you may be wondering how to select a certain number of the names randomly.

As is often the case with Excel, there are a number of different approaches you can take. Each approach examined in this tip assumes that the names you need to select from are listed in cells A1 through A1000. Of course, your range of names could be shorter or longer, but the point is that they are in contiguous cells in column A. The examples also assume that you need to select 15 names at random from the list.

The first approach is to use the INDEX function. Enter the following formula in cells B1:B15:

```
=INDEX(A:A,INT((RAND()*1000)+1),1)
```

A similar formula uses the OFFSET function:

```
=OFFSET($A$1,ROUNDUP(RAND()*1000,0),0,1,1)
```

It is possible, but not probable, that you will get the same name twice in the resulting list. (The improbability comes because of the size of the original list. The larger the list, the less probable there will be duplicates in the extracted list.) If you do get a duplicate name, then simply force a recalculation of your worksheet by pressing **F9**. Each time your recalculate, the list of extracted names is regenerated.

Another potential approach requires the use of multiple columns. Simply follow these steps:

1. Enter **=RAND()** in cell B1.
2. Enter the following formula in cell C1:

```
=RANK(B1,$B$1:$B$1000)
```

3. Select the range B1:C1, and fill down to row 1000.
4. Select the range B1:C1000.
5. Press **Ctrl+C** to copy the range to the Clipboard.
6. Display the Paste Special dialog box. (In Excel 2007 or Excel 2010, display the Home tab of the ribbon and click the down-arrow under the Paste tool at the left side of the ribbon. Select Paste Special from the resulting menu. In older versions of Excel choose Paste Special from the Edit menu.)

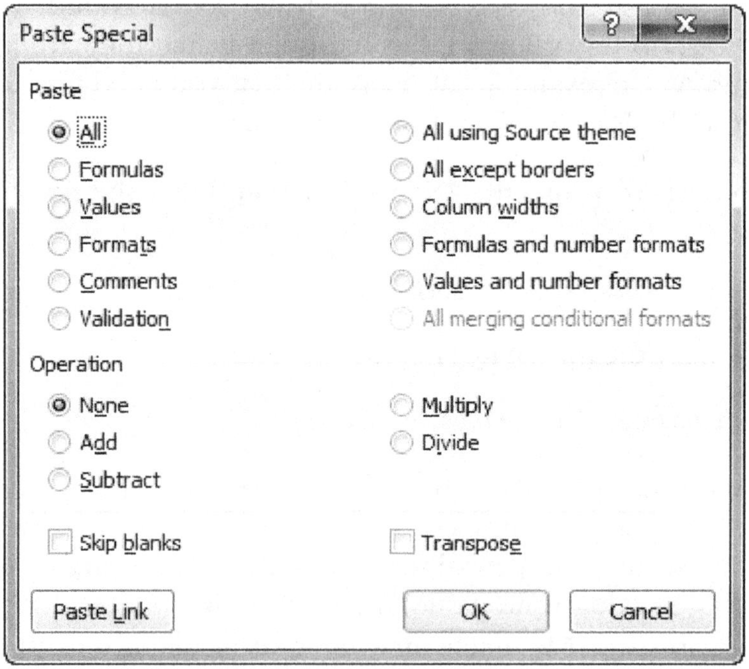

The Paste Special dialog box.

7. Make sure the Values radio button is selected.
8. Click on OK. You now have static values in B1:C1000, which means they won't change every time the worksheet is recalculated.
9. Select a cell in column C.
10. Display the Sort dialog box. (In Excel 2007 or Excel 2010 display the Data tab of the ribbon and then click Sort. In older versions of Excel Sort from the Data menu.)

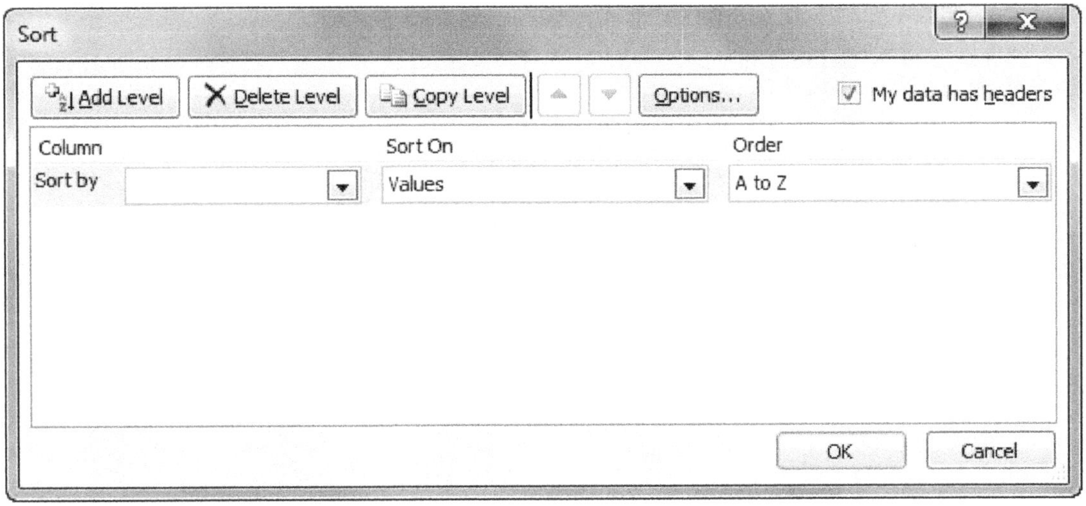

The Sort dialog box.

11. Click on OK. The table (range A1:C1000) is sorted according to the values in column C.

The result is that column C now contains a ranking of all the random numbers in column B. The first 15 rows contain your random names.

In this approach you could also have left out column C completely and simply sorted your list based on the static random values in column B. Again, the top 15 would be your random names.

Of course, there are any number of macro solutions you could use for this problem. The coding of any macro will be similar, relying on VBA's RND function to generate random numbers. Of all the possible macro solutions, perhaps the following is the most unique and offers some advantages not available with the workbook solutions discussed so far:

```
Sub GetRandom()
    Dim iRows As Integer
    Dim iCols As Integer
    Dim iBegRow As Integer
    Dim iBegCol As Integer
    Dim J As Integer
    Dim sCells As String

    Set TempDO = New DataObject
```

```
        iRows = Selection.Rows.Count
        iCols = Selection.Columns.Count
        iBegRow = Selection.Row
        iBegCol = Selection.Column

        If iRows < 16 Or iCols > 1 Then
            MsgBox "Too few rows or too many columns"
        Else
            Randomize Timer
            sCells = ""
            For J = 1 To 15
                iWantRow = Int(Rnd() * iRows) + iBegRow
                sCells = sCells & Cells(iWantRow, iBegCol) & vbCrLf
            Next J
            TempDO.SetText sCells
            TempDO.PutInClipboard
        End If
    End Sub
```

To use this macro, just select the names from which you want to select the 15 random names. In the examples thus far, you would select the range A1:A1000. The macro then pulls 15 names at random from the cells, and puts them in the Clipboard. When you run the macro, you can then paste the contents of the Clipboard where ever you want. Every time the macro is run, a different group of 15 is selected.

Generating Random Strings of Characters

Nancy is trying to get Excel to pick 50 "numbers" that each contain eight random characters. The characters can be either digits or letters (uppercase or lowercase).

If your random numbers were to really be numbers (digits only), then generating them would be easy. All you would need to do is use the RANDBETWEEN function (part of the Analysis ToolPak before Excel 2007) in this manner:

```
=RANDBETWEEN(10000000,99999999)
```

This is not what Nancy wants, however. Her random "numbers" can contain upper- and lowercase letters, as well. This becomes a bit stickier. There are, however, several approaches you can use.

One approach is to put all your possible characters into an individual cell, such as B7:

```
ABCDEFGHIJKLMNOPQRSTUVWXYZabcdefghijklmnopqrstuvwxyz0123456789
```

Name this cell something snazzy, such as MySource. You could then use a formula such as the following to return the random string of characters:

```
=MID(MySource,RANDBETWEEN(1,LEN(MySource)),1)
& MID(MySource,RANDBETWEEN(1,LEN(MySource)),1)
& MID(MySource,RANDBETWEEN(1,LEN(MySource)),1)
& MID(MySource,RANDBETWEEN(1,LEN(MySource)),1)
& MID(MySource,RANDBETWEEN(1,LEN(MySource)),1)
& MID(MySource,RANDBETWEEN(1,LEN(MySource)),1)
& MID(MySource,RANDBETWEEN(1,LEN(MySource)),1)
& MID(MySource,RANDBETWEEN(1,LEN(MySource)),1)
```

The formula is long; it has been broken into individual lines for clarity, but it is still a single formula. It concatenates eight characters pulled from the source you entered into cell B7.

Another approach is to create a table that contains all the characters you would want in your random text string. Start by placing the numbers 1 through 62 in a column, one number in each row. To the left of these numbers place your characters—A, B, C, D, etc. (This should be the same characters you placed in cell B7 in the previous technique.) Select both columns of the 62 rows and give it a name, such as MyTable. You can then use the following formula to generate the random characters:

```
=VLOOKUP(RANDBETWEEN(1,62),MyTable,2)
& VLOOKUP(RANDBETWEEN(1,62),MyTable,2)
& VLOOKUP(RANDBETWEEN(1,62),MyTable,2)
& VLOOKUP(RANDBETWEEN(1,62),MyTable,2)
& VLOOKUP(RANDBETWEEN(1,62),MyTable,2)
& VLOOKUP(RANDBETWEEN(1,62),MyTable,2)
& VLOOKUP(RANDBETWEEN(1,62),MyTable,2)
& VLOOKUP(RANDBETWEEN(1,62),MyTable,2)
```

Again, remember that this is a single formula, although it is a bit shorter than the previous formula.

Each of the approaches presented so far has one drawback: they are regenerated each time your worksheet is recalculated. Thus, it is hard to have a single generated random string that won't change on a regular basis. The best way around this is to use a macro, but you don't necessarily want to use a user-defined function. Why? Because it, too, would change its result every time the worksheet was recalculated. Instead, you need a macro that will put the random strings into your workbook starting a a specific cell location. The following is an example of such a macro:

```
Sub MakeRandom()
    Dim J As Integer
    Dim K As Integer
    Dim iTemp As Integer
    Dim sNumber As String
    Dim bOK As Boolean

    Range("D4").Activate
    Randomize
    For J = 1 To 50
        sNumber = ""
        For K = 1 To 8
            Do
                iTemp = Int((122 - 48 + 1) * Rnd + 48)
                Select Case iTemp
                    Case 48 To 57, 65 To 90, 97 To 122
                        bOK = True
                    Case Else
                        bOK = False
                End Select
            Loop Until bOK
            bOK = False
            sNumber = sNumber & Chr(iTemp)
        Next K
        ActiveCell.Value = sNumber
        ActiveCell.Offset(1, 0).Select
    Next J
End Sub
```

Run the macro, and whatever is in cells D4:D53 is overwritten by the random values. If you want the values written into a different location, change the Range statement near the beginning of the macro.

7

Calculations Involving Lookups

Calculating the Interval between Occurrences

Roger asked if there was a way to calculate the interval between occurrences of values in a list. For instance, he has several thousand numbers in column A. Looking at the value in cell A351, the last time that value occurred in the list was in cell A246. He would like a formula that could be placed in cell B351 and return 105, the difference between 351 and 246.

This approach is difficult to implement in Excel because Excel is not very good at searching backwards—up a column. If the premise could be reversed, then the task becomes much simpler. For instance, if a formula in cell B246 could return the value 105, indicating the interval until the next occurrence of the value in cell A246, instead of calculating the last occurrence. The following formula calculates the next occurrence of the value in cell A1:

```
=MATCH(A1,A2:$A$65536,0)
```

Place this formula in cell B1 and copy it down however many cells are necessary. If the value in column A does not occur again in the column, then the formula returns the #N/A error. If you would rather have the formula return 0, then the following works:

```
=IF(ISNA(MATCH(A1,A2:$A$65536,0)),0,MATCH(A1,A2:$A$65536,0))
```

Remember, in the above examples, that you may need to adjust the row indicator (65536) to something that reflects the largest row number you anticipate. This is particularly true if you are using Excel 2007 or Excel 2010, as their worksheets can contain many more rows than was the case in earlier versions of Excel.

If you absolutely must count upwards (find the previous occurrence instead of the next occurrence), then the easiest way to do it is with a user-defined function. The following function, RowInterval, will look backward through a range you specify and return the desired interval:

```
Function RowInterval(TestCell As Range, LookHere As Range) As Long
    Dim varValue As Variant
    Dim lngRow As Long

    Application.Volatile
    varValue = TestCell.Value

    'Check for occurrences of the test value in the search range
    If WorksheetFunction.CountIf(LookHere, varValue) > 0 Then
        With LookHere
            'Get the last row of the search range
            lngRow = .Row + .Rows.Count - 1
            'Start with the last cell in the search range and work up
            Do Until .Item(lngRow, 1).Value = varValue
                lngRow = lngRow - 1
            Loop
        End With
        'Subtract the number of the row containing the found occurrence
        'from the number of the row containing the test value
        RowInterval = TestCell.Row - lngRow
    End If
End Function
```

In order to use the function, you would put the following formula in cell B2, and then copy the formula down the number of desired cells:

```
=RowInterval(A2,A$1:A1)
```

Inserting Dashes between Letters and Numbers

Let's say you have a worksheet with lots of product codes in column A. These codes are in the format A4, B12, AD4, etc. Due to a change in the way your company operates, you are directed to change all the product codes so they contain a dash between the letters and the numbers.

There are several ways you can perform this task. If the structure of your product codes is consistent, then inserting the dashes is a snap. For instance, if there will always be a single letter followed by numbers, then you could use a formula such as this:

```
=LEFT(A1,1) & "-" & RIGHT(A1,LEN(A1)-1)
```

Chances are good that your data won't be structured, and that you could have one or two letters followed by up to three digits. Thus, both A4 and QD284 would both be valid product codes. In this case, a solution formula takes a bit more creativity.

One way to handle it is with an array formula. Consider the following formula:

```
=REPLACE(A1,MATCH(FALSE,ISERROR(1*MID(A1,ROW(INDIRECT("1:100")),1)),0),0,"-")
```

If values are in A1-A10, you can put this formula into B1, and then copy it down the column. Since it is an array formula, it must be entered by pressing **Ctrl+Shift+Enter**. The formula finds the location of the first number in the cell, and inserts a dash before it.

Assume, for the sake of example, that cell A1 contains BR27. The innermost part of the formula, INDIRECT("1:100"), converts the text 1:100 to a range. This is used so that inserting or deleting rows does not affect the formula. The next part of the formula, ROW(INDIRECT("1:100")), essentially creates an array of the values 1-100: 1,2,3,...,99,100. This is used to act on each character in the cell.

The next part, MID(A1,ROW(INDIRECT("1:100")),1), refers to each individual character in the string. This results in the array: "B", "R", "2", and "7". Multiplying the array by 1 (the next part of the formula) results in each of the individual characters being converted to a number. If the character is not a number, this conversion yields an error. In the case of the string being converted (BR27), this results in: #VALUE, #VALUE, 2, and 7.

The next step is to apply the ISERROR function to the results of the multiplication. This converts the errors to TRUE and the non-errors to FALSE, yielding TRUE, TRUE, FALSE, and FALSE. The MATCH function looks in the array of TRUE and FALSE values for an exact match of FALSE. In this example, the MATCH function returns the number 3, since the first FALSE value is in the third position of the array. At this point, we essentially know the location of the first number in the cell.

The final function is REPLACE, which is used to actually insert the dash into the source string, beginning at the third character.

As you can tell, the formula to perform the transformation can be a bit daunting to decipher. For those so inclined, it may be easier to just create a user-defined function. The following macro is an example of one that will return a string with the dash in the proper place:

```
Function DashIn(myText As String)
    Dim i As Integer
    Dim myCharCode As Integer
    Dim myLength As Integer

    Application.Volatile
    myLength = Len(myText)
    For i = 1 To myLength
        myCharCode = Asc(Mid(myText, i, 1))
        If myCharCode >= 48 And myCharCode <= 57 Then
            Exit For
        End If
    Next i
    If i = 1 Or i > myLength Then
        DashIn = myText
    Else
        DashIn = Left(myText, i - 1) & "-" _
          & Mid(myText, i, myLength - 1)
    End If
End Function
```

The macro examines each character in the original string, and when it finds the first numeric character, it inserts a dash at that point. You would use the function in this way:

```
=DashIn(A1)
```

Breaking Up Variable-Length Part Numbers

Marty has a worksheet with a long series of part numbers in column A. These consist of letters and numbers, such as A123BC, AB123C, etc. Marty wants to break the data into three columns, so that text before the numbers will be in one column, numbers in the second column, and text after the numbers in the third.

The factor that complicates dividing the part number into segments is that there is no set length for each component of the combined part number. If the components were of standard lengths, then you could use the Text to Columns function in Excel. Since they aren't, and there is no delimiter between the components, then that potential avenue for solution is closed.

If you want to use formulas to pull apart the part numbers, then you will need three of them, one for each component you want to extract. Assuming that the part numbers follow the pattern indicated (text, digits, text) and that the first part number is in cell A1, you could use the following in cell B1:

```
=LEFT(A1,MATCH(FALSE,ISERROR(1*MID(A1,ROW(INDIRECT("1:100")),1)),0)-1)
```

This needs to be entered as an array formula, meaning that you need to enter it using **Ctrl+Shift+Enter**. The formula finds the first numeric digit in the part number, and then returns everything before that digit. It will work on any part number that isn't over 100 characters in length.

To extract the second component of the part number, you can put the following formula in cell C1:

```
=MID(A1,MATCH(FALSE,ISERROR(1*MID(A1,ROW(INDIRECT("1:100")),
1)),0),COUNT(1*MID(A1,ROW(INDIRECT("1:100")),1)))
```

Again, this is a single formula, and it needs to be entered as an array formula (**Ctrl+Shift+Enter**) so that it can work on each character in the original part number. It examines the part number and determines the beginning point of the digits, and then extracts all those digits. It returns a text string, even though that string is composed of digits. If you want it to actually be treated as a number (which would get rid of any leading zeros, of course), then you need to enclose the entire formula in a Value function, as shown here:

```
=VALUE(MID(A1,MATCH(FALSE,ISERROR(1*MID(A1,ROW(INDIRECT("1:100")),
1)),0),COUNT(1*MID(A1,ROW(INDIRECT("1:100")),1))))
```

To get the last component of the part number, you need to use the following formula, again entered as an array formula:

```
=MID(A1,MATCH(FALSE,ISERROR(1*MID(A1,ROW(INDIRECT("1:100")),
1)),0)+COUNT(1*MID(A1,ROW(INDIRECT("1:100")),1)),100)
```

While this approach works very well, array formulas are notoriously calculation intensive, especially when you have a lot of the formulas in your worksheet. If you need to pull apart a thousand part numbers, that means that you end up with 3,000 array formulas, which can be very, very slow in recalculating.

If you find yourself in this situation, you may want to use a macro to actually pull apart the part numbers. The following macro should work on part numbers that follow the pattern of text, digits, text, as already described.

```
Sub Split1()
    Dim C As Range
    Dim sNew As New
    Dim i As Integer

    For Each C In Selection
        sNew = ""
        i = 1

        ' Get first part, which is text
        Do While IsNumeric(Mid(C, i, 1)) = False
            sNew = sNew & Mid(C, i, 1)

            i = i + 1
            If i > Len(C) Then Exit Do
        Loop
        C.Offset(0, 1).Value = sNew

        ' Pull next part, which should be digits
        sNew = ""
        Do While IsNumeric(Mid(C, i, 1)) = True
            sNew = sNew & Mid(C, i, 1)

            i = i + 1
            If i > Len(C) Then Exit Do
        Loop
        C.Offset(0, 2).Value = sNew

        ' Use rest of original cell
        sNew = Mid(C, i, Len(C))
        C.Offset(0, 3).Value = sNew
    Next C
End Sub
```

To use the macro, just make a selection of part numbers and run it. The macro uses the concept of looking for changes between numeric/nonnumeric values in string of characters in the cell. When one of these boundaries is reached, the part of the original string before the boundary is stuffed into a new cell. This concept can be shortened a bit, as is done in the following example.

```
Sub Split2()
    Dim C As Range
    Dim j As Integer
    Dim k As Integer

    For Each C In Selection
        j = 1
        Do While Not (IsNumeric(Mid(C.Value, j, 1))) And j <= Len(C)
            j = j + 1
        Loop
        k = j

        Do While IsNumeric(Mid(C.Value, k, 1)) And k <= Len(C)
            k = k + 1
        Loop

        C.Offset(0, 1) = Left(C, j - 1)
        C.Offset(0, 2) = Mid(C, j, k - j)
        C.Offset(0, 3) = Mid(C, k, Len(C) - (k - 1))
    Next C
End Sub
```

The difference between this version of the macro and the previous one, of course, is that this version steps through the original cell and determines the boundaries all at once. When they are known, then the components of the original part number are stuffed into the cells.

An interesting approach to pulling apart the part numbers is to use a couple of short user-defined functions that determine where the boundaries are between the components. Consider the following two functions:

```
Function pNumber(X)
    i = 1
    Do Until Mid(X, i, 1) Like "#": i = i + 1: Loop
    pNumber = i
End Function

Function pAlpha(X)
    X = UCase(X)
    i = pNumber(X)
    Do Until Mid(X, i, 1) Like "[A-Z]": i = i + 1: Loop
```

```
    pAlpha = i
End Function
```

These are much shorter than the previous macros, and all they do is return the boundary where the numbers start (in the case of pNumber) and the boundary where the second group of text starts (in the case of pAlpha). To use the functions, you use the following three formulas to return, respectively, the first, second, and third components of the original part number:

```
=MID(A1,1,pNumber(A1)-1)
=MID(A1,pNumber(A1),pAlpha(A1)-pNumber(A1))
=MID(A1,pAlpha(A1),LEN(A1)-pAlpha(A1)+1)
```

Finding the First Non-Digit in a Text Value

Tony has a bunch of data in a worksheet that consists of digits and other characters. For instance, he might have a cell that contains "1234567Blue." Tony wants to be able to figure out the character position at which the first non-digit character occurs. In the example of the text "1234567Blue" Tony wants some way to figure out that the first non-digit character is at position 8.

There are two primary ways to get the value you want. The first is to use an array formula to calculate the position. The following array formula (entered by using **Ctrl+Shift+Enter**) will work in the majority of cases:

```
=MATCH(TRUE,ISERROR(VALUE(MID(A1,ROW(INDIRECT("1:"&LEN(A1))),1))),0)
```

The only instances where this array formula won't work is if cell A1 is either empty or contains a strictly numeric value. If your list may contain this type of data (or no data at all), then you should consider using a slightly longer array formula:

```
=IF(LEN(A1)=0,0,MIN(IF(1*ISNUMBER(1*MID(A1,ROW(INDIRECT("A1:A"&
LEN(A1))),1))=0,ROW(INDIRECT("A1:A"&LEN(A1))),LEN(A1)+1)))*
(ISNUMBER(A1)=FALSE)
```

Remember that that is a single array formula, entered by using **Ctrl+Shift+Enter**. It will properly handle instances where A1 contains no non-digit characters (as in a blank cell or a value such as "123").

Of course, the other way you can handle finding out the position of the first non-digit character is to create a user-defined function. There are many different ways that such a macro can be implemented. One of the easiest ways to implement the macro is to simply step through each character in whatever is passed to the macro. When a character is found that is outside the ASCII code range for digits (48 to 57), then you know you've found the first position. The following macro shows a way to do this type of technique:

```
Function FirstNonDigit(str As String)
    Dim iChar As Integer
    Dim iPos As Integer
    Dim J As Integer

    Application.Volatile
    iPos = 0
    For J = 1 To Len(str)
        iChar = Asc(Mid(str, J, 1))
        If iChar <= 47 Or iChar >= 58 Then
            iPos = J
            Exit For
        End If
    Next J
    FirstNonDigit = iPos
End Function
```

To use the function, simply use a formula such as this in your worksheet:

```
=FirstNonDigit(A1)
```

If the cell you reference is empty or if it only contains digits, then the function returns a 0 value.

Calculating the Median Age of a Group of People

Jan has a worksheet that has a list of ages in column A. The ages start at 1 and go through to 100. In column B she has the number of people in each of those ages. Jan needs a formula that will tell her the median age of this group of people.

At first blush you might think that you can use the MEDIAN function to calculate the median. That function works great if you have a simple list of values. For instance, if you were calculating the median for the ages alone, then MEDIAN would work fine. However, in Jan's case she needs the median value for the ages of the people, not for the ages themselves. In other words, the median needs to be weighted by the number of people that are each age. The MEDIAN function cannot handle such a requirement.

It should be pointed out that the median age is going to be different than the average age for a group of people. The average can be calculated most easily by multiplying the age by the number of people that is each age. For instance, in column C you could place a formula such as =A1*B1 and then copy it down the column. Add up the values in columns B and C, and then divide the sum in column C by the sum in column B. The result is the average age for the list of people.

The median age, on the other hand, is the age at which half of the people fall below that age and half above that age. The median age can best be calculated by using an array formula, such as the following:

```
=MATCH(SUM($B$1:$B$100)/2,SUMIF($A$1:$A$100,
"<="&$A$1:$A$100,$B$1:$B$100))
```

This is a single formula, entered by pressing **Ctrl+Shift+Enter**. The SUMIF function in the formula is used to generate an array of the cumulative number of people who are less than or equal to each age. The SUM portion of the formula gives the midpoint of the total frequency of ages. The MATCH function is then used to look up the midpoint value in the array of cumulative frequencies. This yields an "index number" in the initial array, and since the array consists of the all ages 1 through 100, this index number matches is equivalent to the median age.

Determining a Simple Moving Average

Jeff needs to create a formula that will return a moving average for a range of cells. He adds data to the worksheet daily and he always want to have an average of the last ten days' information. This always corresponds to the last ten cells in a column.

There are a couple of easy ways you can approach this problem. The solution you choose depends on what you ultimately want to see in the way of an average. For instance, if you want to see how the average changes over time, the best approach is to add an additional column to your worksheet. If the data is in column A (starting in row 2), then you can enter the following formula in cell B11:

```
=IF(A11>"",AVERAGE(A2:A11),"")
```

Copy the formula down the column, and you will always have the average of the last ten days shown. As you add new data to column A, the updated moving average appears at the bottom of column B. The advantage is that you can see how the average changes from day to day.

If you don't want to add another column for each day's moving average, you can use a different formula to determine the current moving average. Assuming there are no blanks in column A and that there are more than ten pieces of data in the column, you could use the following formula:

```
=AVERAGE(OFFSET(A1,COUNTA(A:A)-1,0,-10,1))
```

The OFFSET function defines the range to average. It looks at the number of cells in column A and selects the last 10 as the desired range.

Averaging the Last Numbers in a Column

Emma has a list of numbers in a worksheet (let's say in column A) that are added to on a weekly basis. She needs to calculate the average of the last 12 numbers in the column. She wonders how she can do this and have the average always reflect the last 12 numbers, even when she keeps adding numbers each week.

Assuming that there are no gaps in your range of numbers, you can calculate the average of the last 12 numbers with this formula:

```
=AVERAGE(OFFSET(A1,COUNTA(A:A)-12,0,12,1))
```

This formula should, of course, be placed in some cell that is not in column A. It uses the COUNTA function to figure out how many cells contain something in column A. If there are 100 cells in use in column A, this means that you end up with a formula being evaluated in this way:

```
=AVERAGE(OFFSET(A1,100-12,0,12,1))
```

Of course, 100 minus 12 is 88, and this number is used as an offset from the starting cell (A2) to say that the range to be averaged should start at A89 and extend down 12 cells. That means that the average ends up being for the range A89:A100. As more numbers are added at the bottom of column A, the formula always reflects the last 12 numbers.

The formula will return an error if column A has fewer than 12 rows worth of data in it. To accommodate that possibility, you may want to alter the formula just a bit:

```
=AVERAGE(OFFSET(A1,COUNTA(A:A)-MIN(COUNTA(A:A),12),0,MIN(COUNTA(A:A),12),1))
```

Instead of using a hard-and-fast value of 12 rows, the MIN function (in two places) returns the minimum of either the actual number of rows or 12. So, if your worksheet only has numbers in cells A1:A5, the MIN function would ensure that the formula only averaged those 5 values.

Referencing the Last Six Items in a Formula

Darryl needs to find the average of the last six entries in a column. The number of items in the column can vary over time as more information is added, but he always wants the average of those last six items.

There are a number of different formulas you can use, probably too many to go over here in detail. With that in mind, it is instructive to look at two particular formulas. The one you choose to use will depend on the characteristics of the data in the column. If there are no blank cells in the column, then finding the average can be done with a relatively simple formula:

```
=AVERAGE(OFFSET(A1,COUNT(A:A)-6,0,6,1))
```

This formula uses the OFFSET function to calculate the proper cells to examine, at the bottom of the column. This formula won't work if there are blank cells in the column. In that case you will need to use a formula that examines the contents of each cell and determines, as part of the calculation process, whether it is blank or not. Array formulas or regular formulas using array functions are great for this purpose. The following example uses the SUMPRODUCT function to accomplish the task:

```
=SUMPRODUCT((A1:A30*((MAX(ROW(A1:A30)*(A1:A30<>""))-ROW(A1:A30))<6))/6)
```

This formula assumes that the cells to be evaluated are in the range of A1:A30; it doesn't matter if there are blank cells in this range. The ROW functions are used to create arrays that determine if the individual cells contain values or not. Only those rows containing values end up being counted, and those are divided by 6 and summed, providing the desired average.

Engineering Calculations

In an engineering environment, it is not unusual to need to "normalize" numbers in some manner. For instance, you may need to show numeric values normalized to multiples of 10^3, such that 7340 is expressed as 7.34 and 73400 is expressed as 73.4.

It is possible in Excel to use a custom number format to express information in scientific notation that will normalize the display of a number to a multiple of 10^3. To do this, you would follow these steps:

1. Select the cells you want formatted.
2. Display the Format Cells dialog box. (In Excel 2007 or Excel 2010 display the Home tab of the ribbon and click the small icon at the lower-right corner of the Number group. In older versions of Excel choose Cells from the Format menu.)
3. Make sure the Number tab is selected.

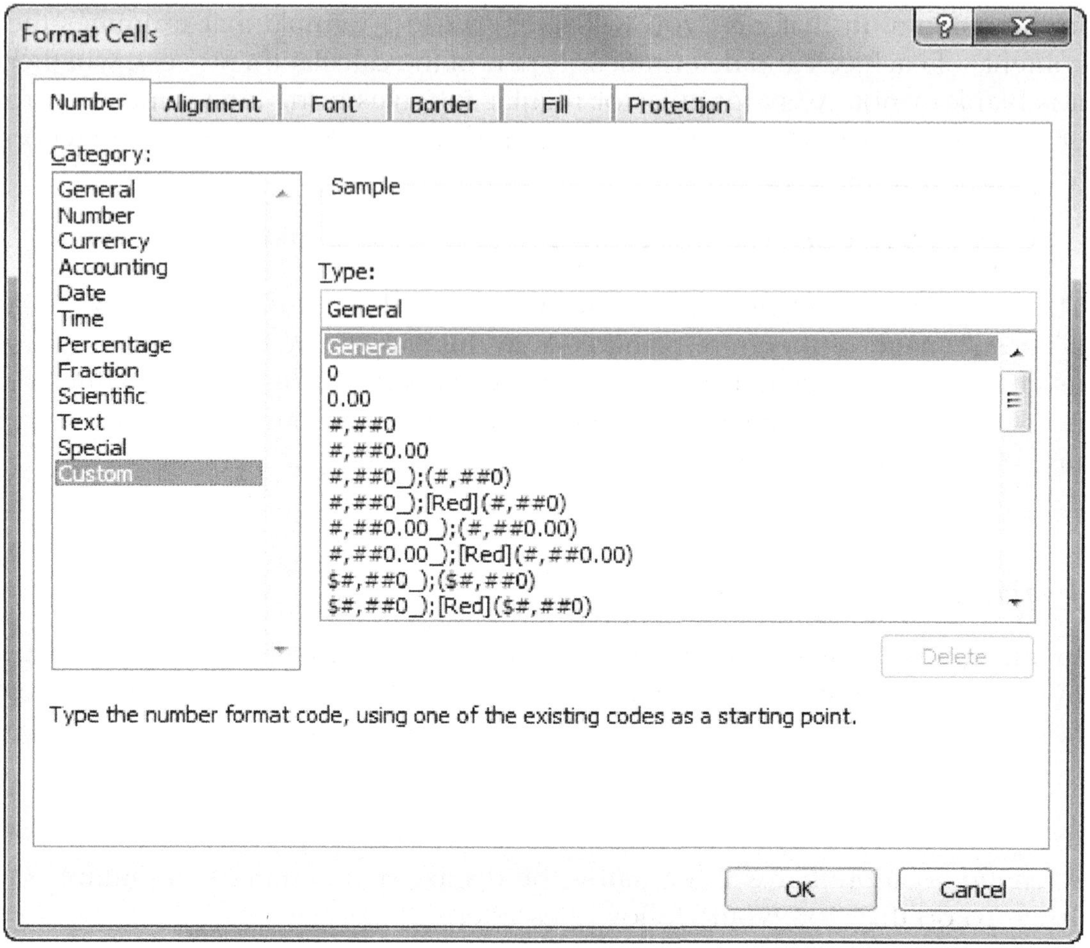

The Number tab of the Format Cells dialog box.

4. In the list of format categories, choose Custom.
5. In the Type box, enter **##0.0E+0** as your format. (This provides only one number to the right of the decimal place. If you want more, increase the number of zeros after the decimal place.)
6. Click on OK.

Now, when you enter a number such as 7340 into the cell, Excel displays it as 7.3E+3. Because of the way the cell format was entered, the portion after the E will always be a multiple of 3.

This is fine and good, but what if you want just the 7.3 in the cell, and then a metric prefix with a unit in an adjoining cell, such as kilograms? This is a bit more complex, but it can be done using formulas. For instance, let's assume you have your original number in cell A2, you wanted the normalized number in cell B2, and the metic prefix and unit name in cell C2. All you would need to do is enter the following formula in cell B2:

```
=IF(OR(A2>=1,A2<=-1),SIGN(A2)*(ABS(A2)/(10^(3*INT(LOG(ABS(A2))/3)))),
IF(A2=0,0,SIGN(A2)*(ABS(A2)*10^(-3*INT(LOG(ABS(A2))/3)))))
```

Assuming the units you are working with are an imaginary unit called a foo, in cell C2 you would use a different formula, as follows:

```
=IF(OR(A2>=1, A2<=-1),CHOOSE(INT(LOG(ABS(A2))/3)+1, "Foos", "Kilofoos",
"Megafoos", "Gigafoos", "Terafoos", "Petafoos", "Exafoos"),
IF(A2=0,"",CHOOSE(INT(-LOG(ABS(A2))/3)+1, "Millifoos", "Microfoos",
"Nanofoos", "Picofoos", "Femtofoos", "Attofoos")))
```

These formulas may seem a bit long, and they are. However, they will work for any number between approximately -9.99999E-18 to 9.99999E+20. For instance, if you put the number .000125 in cell A2, then cell B2 will contain 125 and cell C2 would contain Millifoos.

If you prefer to not use longer formulas such as these in your workbooks, you can develop a couple of VBA functions to do the trick. The following function, MySciNum, returns a normalized number. Thus, you would use =MySciNum(A2) in cell B2 to get the same results as noted above:

```
Function MySciNum(BaseNum As Double) As Double
    Select Case BaseNum
        Case Is >= 1
            While Abs(BaseNum) > 1000
                BaseNum = BaseNum / 1000
            Wend
        Case 0
            'Do nothing
        Case Else
            While Abs(BaseNum) < 1
                BaseNum = BaseNum * 1000
            Wend
    End Select
    MySciNum = BaseNum
End Function
```

This function only returns a number. To return the units with the appropriate metric prefix, you would use the following function. All you need to do is pass it the cell reference and the name of a single unit. For instance, you could use =MySciPre(A2, "foo"). The macro is as follows:

```
Function MySciPre(BaseNum As Double, Unit As String) As String
    Dim OrigNum As Double
    Dim Pref As Integer
    Dim Temp As String

    Pref = 0
    OrigNum = BaseNum
    Select Case BaseNum
        Case Is >= 1
            While Abs(BaseNum) > 1000
                BaseNum = BaseNum / 1000
                Pref = Pref + 1
            Wend
        Case 0
            Pref = 99
        Case Else
            While Abs(BaseNum) < 1
                BaseNum = BaseNum * 1000
                Pref = Pref - 1
            Wend
    End Select

    Select Case Pref
        Case -6
            Temp = "atto" & Unit
        Case -5
            Temp = "femto" & Unit
        Case -4
            Temp = "pico" & Unit
        Case -3
            Temp = "nano" & Unit
        Case -2
            Temp = "micro" & Unit
        Case -1
            Temp = "milli" & Unit
        Case 0
            Temp = Unit
        Case 1
            Temp = "kilo" & Unit
        Case 2
            Temp = "mega" & Unit
```

```
        Case 3
            Temp = "giga" & Unit
        Case 4
            Temp = "tera" & Unit
        Case 5
            Temp = "peta" & Unit
        Case 6
            Temp = "exa" & Unit
        Case Else
            Temp = ""
    End Select

    If Len(Temp) > 0 Then
        Temp = LCase(Temp)
        Temp = UCase(Left(Temp, 1)) & Mid(Temp, 2)
        If Abs(OrigNum) <> 1 Then Temp = Temp & "s"
    End If

    MySciPre = Temp
End Function
```

8

Adjusting the Way VLOOKUP Works

Making VLOOKUP Trigger a Macro

Mike uses VLOOKUP regularly in his worksheets, but wonders if there is a way to make the function run a macro if it fails to return a value.

There are a couple of ways you could approach this problem. First, you could use a conditional formula to determine whether VLOOKUP will return a value or an error. If it will return an error, then you can have the formula run a user-defined function (MyUDF), as shown here:

```
=IF(ISERROR(VLOOKUP(B2,CODES,1,FALSE)),MyUDF(),
VLOOKUP(B2,CODES,1,FALSE))
```

All you need to do is make sure that you put your actual VLOOKUP code in the formula (twice) and replace MyUDF with the name of the user-defined function you want to trigger.

Another approach is to set up an event handler for the Calculate event. This can be rather simple, as in the following:

```
Private Sub Worksheet_Calculate()
   If IsError(Range("A1")) Then Call Macro1
End Sub
```

This example assumes that the VLOOKUP formula is in cell A1 and that you want to run a macro called Macro1 if the VLOOKUP returns an error. Your macro could then do whatever you need it to do. Remember, as well, that the Calculate event handler should be placed in the ThisWorksheet object.

You could also make the Calculate event handler a bit more robust, as shown here:

```
Private Sub Worksheet_Calculate()
    On Error GoTo myMac
    Worksheets(1).Select
    If Range("A1").Value Then
        Exit Sub
    End If
myMac:
    Macro1    'macro to run if VLOOKUP fails
End Sub
```

Adjusting the VLOOKUP Function

John has a lookup table of ascending numerical values in column A and corresponding text values in column B. When he uses the VLOOKUP function it returns the text value for the numerical value equal to or less than the specified lookup value. John really wants the first value equal to or greater than the lookup value.

There is no way to change how VLOOKUP does its work; it will always match to the value equal to or less than the lookup value. An option, though, is to modify the formula used to do the actual lookup. Consider the following formula, which assumes that the value you want to use for your lookup is in cell D1:

```
=IF(VLOOKUP(D1,$A$1:$A$10,1)=D1,VLOOKUP(D1,$A$1:$B$10,2),
INDEX($B$1:$B$10,1+MATCH(D1,$A$1:$A$10)))
```

If the value in D1 is an exact match to a value in column A, then the regular VLOOKUP formula is used. If it isn't, then VLOOKUP is abandoned in favor of the INDEX function in conjunction with the MATCH function.

If you are able to sort your data table in descending order, you can use a shorter formula:

```
=INDEX($A$1:$B$10,MATCH(D1,$A$1:$A$10,-1),2)
```

The MATCH function uses the D1 value to look for the smallest value that is great than or equal to that value. (This is what the -1 parameter specifies.) The MATCH function returns the row number of the proper row, and then this is used by INDEX to actually fetch the value.

Another interesting approach to the problem is to use the inverse of the lookup values as a control column that will be used to actually look up information from the data table. For instance, let's assume that your data table is in A1:B10, with the actual numerical values you use for the lookup in column A. You need to insert a column to the left of your data table. Into the first cell of this new column (now column A), insert the following formula:

```
=1/B1
```

This provides the inverse of the value in B1, and you can copy it down the cells in column A. Your data table now has three columns, A1:C10. Next, sort your data table based on this new column, in ascending order.

Now you make a slight modification to your lookup formula so that it looks up the inverse of what you want. Assuming that the value you want to use for your lookup is in cell E1, you would use the following formula:

```
=VLOOKUP(1/E1,$A$1:$C$10,3)
```

What you effectively end up with is the desired value, from column C, that is associated with the value in column B that is equal to or less than the value in E1.

Making VLOOKUP Case Sensitive

Robin asked if there is a way to do a VLOOKUP that is case sensitive. Her lookup table/range has entries that are similar (AbC and aBC) with the only difference being the case of the letters. She can't change the values (make them all upper or lower case) since the unique values are vital.

The VLOOKUP function doesn't have a way to check for the case of information; it is case insensitive. There are several ways you can work around this shortcoming, however. One way is to use the CODE function to create an intermediate column that can be searched by VLOOKUP. Assuming that your original data is in column B, you could put the following formula in cell A1 and copy it down the column:

```
=CODE(LEFT(B1,1))&"."&CODE(MID(B1,2,1))&"."&CODE(RIGHT(B1,1))
```

This formula looks at the first three characters of whatever is in cell B1 and converts those characters to decimal character codes separated by periods. Thus, if A1 contained "ABC" then B1 would contain "65.66.67". Assuming that the value you want to locate is in cell C1, you could use the following as your VLOOKUP formula:

```
=VLOOKUP(CODE(LEFT(C1,1))&"."&CODE(MID(C1,2,1))&"."&
CODE(MID(C1,3,1)), A:B,2,)
```

Another approach is to use the EXACT function to determine the location of what you are looking for. This approach doesn't use VLOOKUP at all, instead relying on the INDEX function. The formula assumes that the cells you want to compare are in column A and what you want to return is the corresponding cell in column B.

```
=IF(MIN(IF(EXACT(C1,$A$1:$A$100),ROW($A$1:$A$100)))=0,NA(),
INDEX($B$1:$B$100,MIN(IF(EXACT(C1,$A$1:$A$100),ROW($A$1:$A$100)))))
```

This formula needs to be entered as an array formula (**Shift+Ctrl+Enter**). The first part of the formula (the first instance of EXACT) compares C1 (what you are looking for) to each value in the range A1:A100. Since this is an array formula, you end up with, in this case, 100 True/False values depending on whether there is an exact match or not. If there is a match, then the first ROW function returns the row of the match and the INDEX function is used to grab the value from column B in that row.

In some instances you might want to create your own user-defined function that will do the lookup for you. The following is an example of such a macro:

```
Function CaseVLook(compare_value, table_array As Range, _
  Optional col_index As Integer = 1)
    Dim c As Range
    Dim rngColumn1 As Range

    Application.Volatile

    Set rngColumn1 = table_array.Columns(1)
    CaseVLook = "Not Found"

    'Loop first column
    For Each c In rngColumn1.Cells
        If c.Value = compare_value Then
            CaseVLook = c.Offset(0, col_index - 1).Value
            Exit For
        End If
    Next c
End Function
```

To use the macro, simply call the function with the value you want to find (say cell C1), the range whose first column should be searched (such as A:B), and optionally the offset of the column within that range, as here:

```
=CaseVLook(C1,A:B,2)
```

A few additional approaches can be found at the following Knowledge Base article:

```
http://support.microsoft.com/?kbid=214264
```

Concatenating Values from a Variable Number of Cells

Pam has two columns of data. In column A there are simple identifiers, such as A, B, C, etc. In column B there are a series of integer values. She can sort the data by the identifier and, secondarily, by the integer values. Now she wants, in column C, to have a formula that will concatenate all the integer values for a particular identifier. Thus, if A1:A4 all contain the identifier A, then in cell C1 she would like to have all the values in B1:B4 concatenated and divided by commas, such as "11, 17, 19, 25". Since the number of rows for each identifier can be different, Pam isn't sure how to go about the concatenation.

The easiest way to accomplish this is to use a macro, which can be created as a user-defined function. Here's an example:

```
Function CatSame(c As Range) As String
    Application.Volatile
    sTemp = ""
    iCurCol = c.Column
    If iCurCol = 3 Then
        If c.Row = 1 Then
            sLast = ""
        Else
            sLast = c.Offset(-1, -2)
        End If
        If c.Offset(0, -2) <> sLast Then
            J = 0
            Do
                sTemp = sTemp & ", " & c.Offset(J, -1)
                J = J + 1
            Loop While c.Offset(J, -2) = c.Offset(J - 1, -2)
            sTemp = Right(sTemp, Len(sTemp) - 2)
        End If
    End If
    CatSame = sTemp
End Function
```

This function basically takes a value that is passed to it (a cell reference) and verifies that the cell reference is for column C. If it is, then it starts to concatenate values from column B based on the values in column A. It only returns the string of concatenated values if the value is column A is different than the value in the row above it. Assuming your identifiers are in column A and your values to be concatenated are in column B, you could place the following in column C:

```
=CatSame(C1)
```

Copy this down as far as necessary in column C and you end up with exactly what Pam wanted.

A more versatile function would be one that would function somewhat like VLOOKUP, but bring back a concatenated list of values that match whatever you are looking up. Consider the following function:

```
Function VLookupAll(vValue, rngAll As Range, _
  iCol As Integer, Optional sSep As String = ", ")
    Dim rCell As Range
    Dim rng As Range
    On Error GoTo ErrHandler

    Application.Volatile
    Set rng = Intersect(rngAll, rngAll.Columns(1))
    For Each rCell In rng
        If rCell.Value = vValue Then _
          VLookupAll = VLookupAll & sSep & _
          rCell.Offset(0, iCol).Value
    Next rCell

    If VLookupAll = "" Then
        VLookupAll = CVErr(xlErrNA)
    Else
        VLookupAll = Right(VLookupAll, Len(VLookupAll) - Len(sSep))
    End If
ErrHandler:
    If Err.Number <> 0 Then VLookupAll = CVErr(xlErrValue)
End Function
```

This function takes up to four arguments. The first is the value you want to match in your lookup. In Pam's instance, this would be the identifier you want, such as A, B, or C. The second argument is the range of cells in which to look for the matches (column A in this case). The third argument is an offset (from the range in the second argument) that represents the values you want concatenated. You can use the function in this manner:

```
=VLookupAll("B",A1:A99,1)
```

If you want to specify a different delimiter between values, you can do it using the optional fourth argument. For instance, the following returns a string where a dash separates each value:

```
=VLookupAll("B",A1:A99,1,"-")
```

The solutions so far have focused on using macros. The reason for this is relatively simple: There isn't a formula-based solution that can do what Pam needs. Using nested IF statements to evaluate what is in column A won't work well because you are limited in how deeply IF statements can be nested.

You could use a formula and an intermediate result if you don't mind having the concatenated values be at the last instance of an identifier in column A. Start by putting this formula in cell C1:

```
=B1
```

This formula should go into cell C2:

```
=IF(A2=A1,C1 & ", " & B2, B2)
```

Copy this formula down as many rows as necessary. What you end up with is an increasingly long series of concatenated values in column C, with the longest in each run being on the same row as the last sequential identifier in column A. You can then put the following in all the applicable cells of column D:

```
=IF(LEN(C2)>LEN(C1),"",C1)
```

This formula only displays the longest strings from column C, which is what Pam needed to begin with.

9

Using Lookup Functions to Enhance Charts

Easily Changing Chart Data Ranges

Excel is great when it comes to creating charts based on data in a data table. The tools on the Insert tab of the ribbon (Excel 2007 or Excel 2010) or the Chart Wizard (earlier versions of Excel) can quickly identify an entire data table, or you can select a portion of a data table and use the same tools to create a chart based just on that portion.

If you change the data range for your chart quite often, it can get tiresome to continually change the data range reference. For instance, if you have a data table that includes several years worth of data, you may want to view a chart that is based on the first five years of data, and then change the data range so the chart refers to a different subset of the data. Make the changes often enough and you'll start casting about for ways to make the changes easier (and more reliably).

One way to do this is with the use of named ranges and several worksheet functions. Let's say that your chart is embedded on a worksheet, but the worksheet is different than the one where the source data is located. On the same sheet as the chart, create two input cells which will serve as "from" and "to" indicators. Name these two cells something like FromYear and ToYear.

On your data worksheet (the one without the chart; I'll call it "Source Data"), the data is arranged with each year in a separate column and a series of cost factors in each row. Start your table in column F and place your years in row 2. Place the cost factors in column E, starting at row 3. Above the years place a capital letter that is the same as the column letter, and in column D place a number that is the same as the row number of the data.

D	E	F	G	H	I
		F	G	H	I
2		2006	2007	2008	2009
3	Cost1	69%	66.60%	73.90%	74.50%
4	Cost2	53.60%	56.90%	64.90%	64.40%
5	Cost3	59.30%	51.30%	62.70%	61.20%

First phase of data preparation

In this example, the chart that is embedded on the other worksheet is based on the data in the range F2:I5. There is nothing special about the chart, but the changes you are getting ready to make will make it dynamic, and therefore much more useful.

Start by placing the following formula in cell B1:

```
="Trends For " & IF(FromYear=ToYear,FromYear,FromYear & " to " & ToYear)
```

This formula provides a dynamic title that you will later use for your chart. Give cell B1 the name **addrTitle**, then place the following formula in cell B2:

```
="'Source Data'!$" & INDEX($F$1:$I$1,1,MATCH(FromYear,$F$2:$I$2)) & "$"
& D2 & ":$" & INDEX($F$1:$I$1,1,MATCH(ToYear,$F$2:$I$2)) & "$" & D2
```

Copy the formula in B2 to the cells B3:B5. The formula returns address strings that represent the desired ranges for the X-axis values and the data series. The actual ranges returned by the formulas will vary, based on the values you enter in the FromYear and ToYear cells on the other worksheet. To make things clearer you can enter some labels into column A.

	A	B	C	D	E	F	G	H	I
1	Title	Trends For 2007 to 2009				F	G	H	I
2	Xvalues	'Source Data'!$G$2:$I$2		2		2006	2007	2008	2009
3	Cost1	'Source Data'!$G$3:$I$3		3	Cost1	69%	66.60%	73.90%	74.50%
4	Cost2	'Source Data'!$G$4:$I$4		4	Cost2	53.60%	56.90%	64.90%	64.40%
5	Cost3	'Source Data'!$G$5:$I$5		5	Cost3	59.30%	51.30%	62.70%	61.20%

Second phase of data preparation

Now you need to name each of the cells in the range B2:B5. Select B2 and in the Name Box (just above column A) enter the name "addrXVal" (without the quotes). Similarly name B3 as addrCost1, B4 as addrCost2, and B5 as addrCost3.

The next step is to create a couple of named formulas that you can use in creating the charts. Display either the Define Name or New Name dialog box, depending on your version of Excel. In Excel 2007 and Excel 2010 display the Formulas tab of the ribbon and, in the Defined Names group, click Define Name. In older versions of Excel click on Insert | Name | Define.

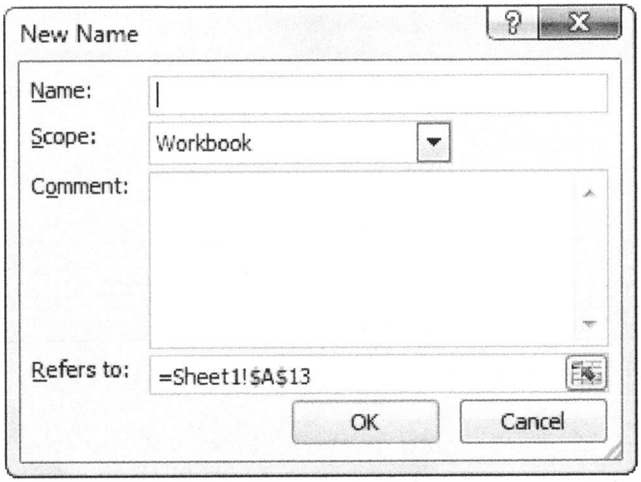

The New Name dialog box.

In the name area type "rngXVal" (without the quotes), then type the following in the Refers To box:

```
=INDIRECT(addrXVal)
```

Click OK and the name is defined. Then, using the same New Name dialog box, define additional names (rngCost1, rngCost2, and rngCost3) that use the same type of INDIRECT formula to refer to the ranges addrCost1, addrCost2, and addrCost3, respectively.

Now you are finally ready to update the references in your chart. How you do this depends on the version of Excel you are using. If you are Excel 2007 or Excel 2010, right-click the chart and choose Select Source. Excel displays the Select Data Source dialog box.

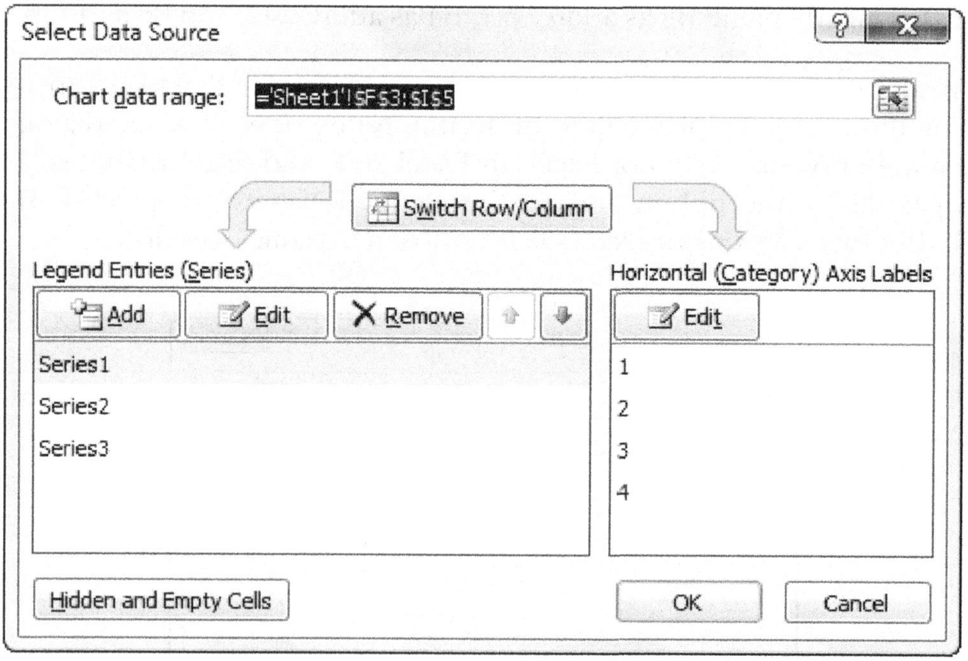

The Series tab of the Source Data dialog box.

For each of the data series listed at the left side of the dialog box, click the Edit button and enter the Name and Values according to the names you defined. Thus, for the Cost1 series you would enter a Name of **='Source Data'!addrCost1** and a Values of **='Source Data'!rngCost1**. You would use the similar references and names for each of the other data series, as well.

Note that you must include the name of your worksheet (Source Data), within apostrophes, in the references you enter. In the Category (X) Axis Labels reference you can enter ='Source Data'!rngXVal.

If you are using an older version of Excel you can update the references in your chart by right-clicking the chart and selecting Source Data, then making sure the Series tab is displayed.

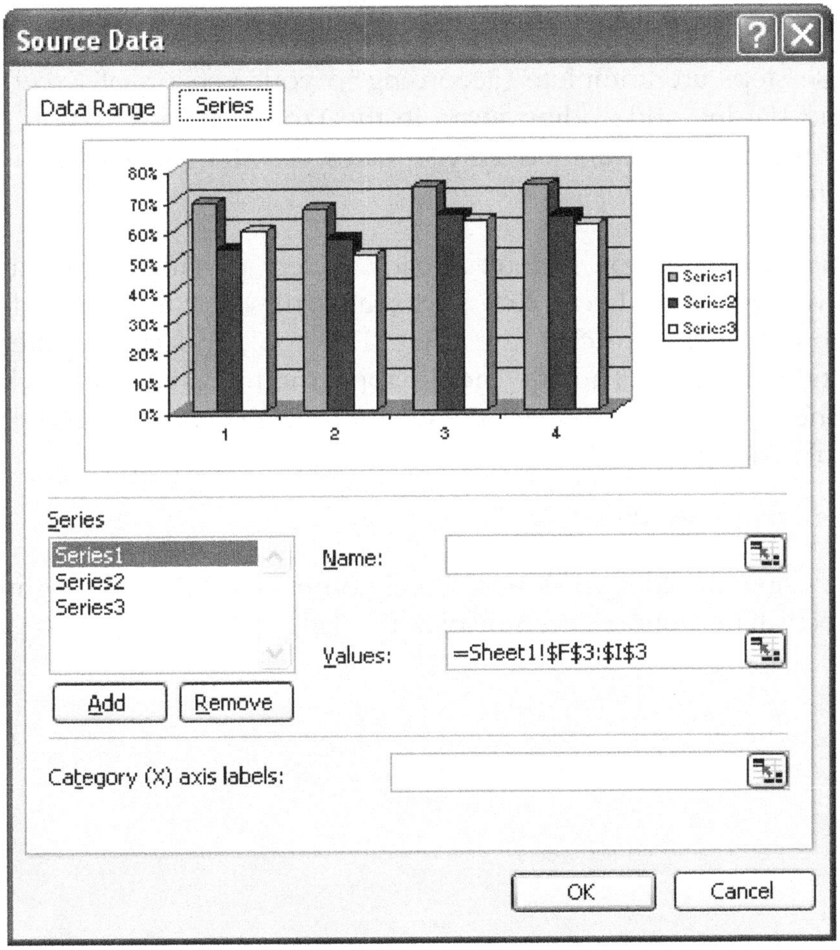

The Series tab of the Source Data dialog box.

For each of the data series listed at the left side of the dialog box, enter the Name and Values according to the names you defined. Thus, for the Cost1 series you would enter a Name of ='Source Data'!addrCost1 and a Values of ='Source Data'!rngCost1. You would use the similar references and names for each of the other data series, as well.

Note that you must include the name of your worksheet (Source Data), within apostrophes, in the references you enter. In the Category (X) Axis Labels reference you can enter ='Source Data'!rngXVal.

Once these steps are complete (according to your version of Excel) you can change the starting and ending years in the FromYear and ToYear cells, and Excel automatically and immediately updates the chart to represent the data you specified.

For an extra touch, if you haven't already added a chart title, go ahead and do so. If you are using Excel 2007 or Excel 2010 select the chart, display the Layout tab of the ribbon, click the Chart Title tool, and choose the way you want the title to appear. The title should appear immediately in the chart. Click the title once to select it. You should see the selection box around the title. In the Formula bar enter the following:

```
='Source Data'!addrTitle
```

If you are using an older version of Excel you can add the title, right-click the chart, select Chart Options, then display the Titles tab.

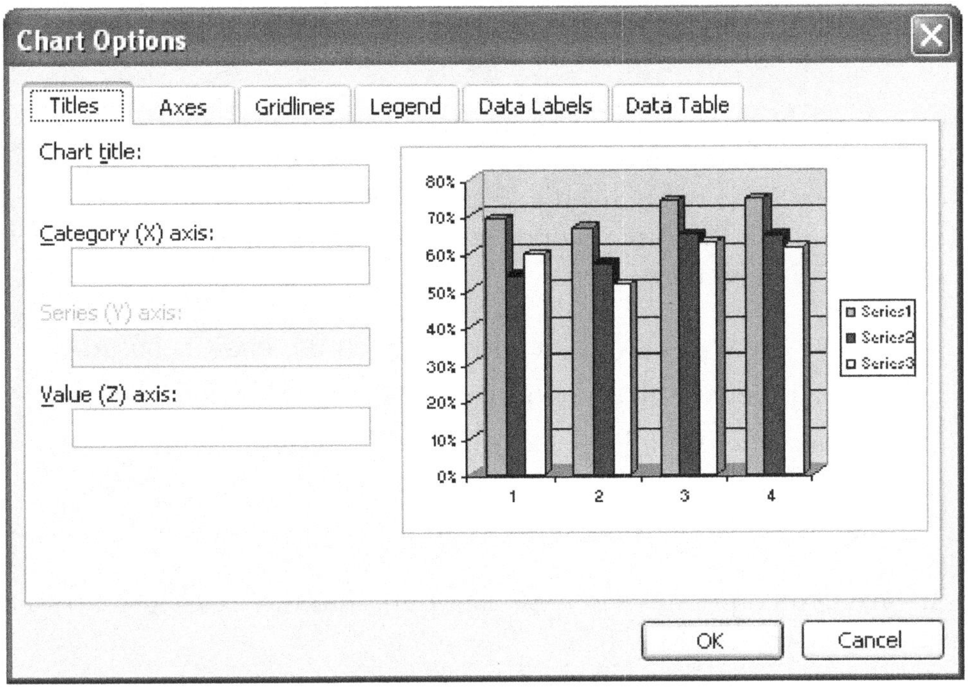

The Titles tab of the Chart Options dialog box.

Enter anything you'd like in the Chart Title box (you'll replace it in a moment), then click OK. The chart title should already be selected, but if it isn't, click on it once. You should see the selection box around the title. In the Formula bar enter the following:

```
='Source Data'!addrTitle
```

The chart title is now linked back to the cell containing the title string, which in turn is dynamically updated each time you change the FromYear and ToYear values.

Automatically Updating Charts for Additional Data

Excel shines at turning your data into charts—graphical representations of your data. You can easily create a chart based on a range of data in a worksheet.

Normally, if you add additional data to your range, you will need to once again create the chart or at best change the range of cells on which the chart is based.

If you get tired of modifying charts to refer to new data ranges, there are a couple of shortcuts you can try out. The first shortcut works fine if you simply need to "fine tune" the range used in a chart. (This approach only works if the chart is on object on the same worksheet that contains the data on which the chart is based.) Follow these steps:

1. Enter any new data to the table on which the chart is based.
2. Select all the cells that you just entered.
3. Click anywhere on the selection border around the cells. (Don't click on the fill handle.)
4. Drag the selection toward the chart and drop it on the chart.

That's it—Excel incorporates the new data right into the existing chart, slick as a whistle.

Another approach is to add new data to the range not at the end, but somewhere within the range. For instance, you may have some data that represents a time period, such as 11/1 through 11/13, and you create a chart based on those dates. If you add new data to the end of the range (after 11/13), then Excel doesn't know you want those items added to the chart.

Instead, insert some blank rows somewhere within the data range; it doesn't matter where, as long as the record for 11/13 is below the added rows. You can then add the new data in the new rows, and the chart is automatically updated to include the data.

One drawback to this approach, of course, is that the inserted data will be out of order when compared to the overall structure of the data table. It is interesting to note that for some types of data—such as those based on dates—Excel will automatically sort the data by date as it presents it in the chart, but not in the data table itself. You can always use Excel's sort feature to reorder the data in the table, all without affecting what is presented in the chart.

Still another approach is to create a "dynamic range." This approach works well if the data range you are charting is the only data on the worksheet. Follow these steps:

1. Create your data table as you normally would. (To keep things simple, we'll assume that you are creating a two-column data table, in columns A and B, that you then want to chart.)
2. Create your chart, as desired.
3. With the worksheet visible, display either the Define Name or New Name dialog box, depending on your version of Excel. In Excel 2007 or Excel 2010 display the Formulas tab of the ribbon and, in the Defined Names group, click Define Name. In older versions of Ecxel click on Insert | Name | Define.

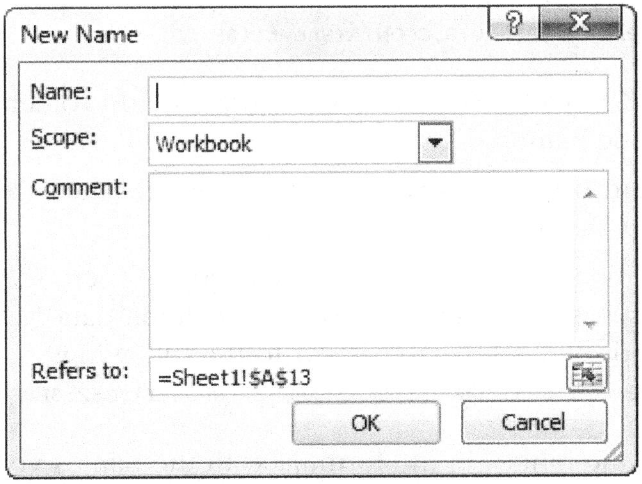

The New Name dialog box.

4. In the Names in Workbook or Name field (depending on your version of Excel) enter a name that will refer to the data in column A.
5. In the Refers To field, enter the following formula. You should change the sheet name and beginning cell (in this case $A$2), as necessary. If you change the beginning cell, you should also change the final part of the formula (-1) to be one less than the row number of the beginning cell. Thus, if you change the beginning cell to $A$5, you should change the final part of the formula to -4.

```
=OFFSET(Sheet1!$A$2,0,0,COUNTA(Sheet1! $A:$A)-1)
```

6. Click OK (Excel 2007 and Excel 2010) or Add (older versions). Excel creates the name and associates it with the formula you entered. In versions of Excel prior to Excel 2007 the Define Name dialog box should still be visible, with the newly defined name listed.
7. If you are using Excel 2007 or Excel 2010, again display the New Name dialog box by displaying the Formulas tab of the ribbon and clicking Define Name.
8. In the Names in Workbook or Name field (depending on your version of Excel) enter a name that will refer to the data in column B.
9. In the Refers To field, enter the following formula. (You should make the same type of changes to the sheet name, beginning cell, and final part of the formula as described in step 5.)

```
=OFFSET(Sheet1!$B$2,0,0,COUNTA(Sheet1!$B:$B)-1)
```

10. Click OK (Excel 2007 and Excel 2010) or Add (older versions). Excel creates the name.
11. If you are using a version of Excel prior to Excel 2007, click OK to close the Define Name dialog box.
12. Display the chart and select the data series. (There should only be one data series, since this is a simple two-column data table.) The formula bar will display a formula similar to the following:

```
=SERIES(Sheet1!$B$1,Sheet1!$A$2:$A$32, Sheet1!$B$2:$B$32,1)
```

13. Replace the ranges in the formula with the names you defined earlier in these steps. Assuming you defined the names ***Dates*** and ***Readings*** in steps 6 and 9, the formula should look like this:

```
=SERIES(,Sheet1!Dates,Sheet1!Readings,1)
```

Now the chart will update automatically regardless of where you add information in your data table. This works because the names you defined in steps 6 and 10 refer to a formula that calculates the extent of data in columns A and B of your worksheet.

There are many other ways that you can create dynamic ranges, depending on the characteristics of the data you are using. For additional information, see these Web resources:

```
http://spreadsheetpage.com/index.php/tip/update_charts_automatically_when_
you_enter_new_data/
http://www.ozgrid.com/Excel/DynamicRanges.htm
```

Automatically Creating Charts for Individual Rows in a Data Table

David has a worksheet that he uses to track sales by company over a number of months. The company names are in column A, and up to fifteen months of sales are in columns B:P. David would like to create a chart that could be dynamically changed to show the sales for a single company from the worksheet.

There are several ways that this can be done; I'll examine three of them in this tip. For the sake of example, let's assume that the worksheet is named MyData, and that the first row contains data headers. The company names are in the range A2:A151, and the sales data for those companies is in B2:P151.

One approach is to use Excel's AutoFilter capabilities. Create your chart as you normally would, making sure that the chart is configured to draw its data series from the rows of the MyData worksheet. You should also place the chart on its own sheet.

Now, select A1 on MyData and apply your filter. In Excel 2007 or Excel 2010 display the Data tab of the ribbon and click the Filter tool. In earlier versions of Excel choose Data | Filter | AutoFilter. A small drop-down arrow appears at the top of each column. Click the drop-down arrow for column A and select the company you want to view in the chart. Excel redraws the chart to include only the single company.

The only potential drawback to the AutoFilter approach is that each company is considered an independent data series, even though only one of them is displayed in the chart. Because they are independent, each company is charted in a different color. If you want the same charting colors to always be used, then you will need to use one of the other approaches.

Another way to approach the problem is through the use of an "intermediate" data table—one that is created dynamically, pulling only the information you want from the larger data table. The chart is then based on the dynamic intermediate table. Follow these steps if you are using Excel 2007 or Excel 2010:

1. Create a new worksheet and name it something like "ChartData".
2. Copy the column headers from the MyData worksheet to the second row on the ChartData sheet. (In other words, copy MyData!A1:P1 to ChartData!A2:P2. This leaves the first row of the ChartData sheet temporarily empty.)
3. With the MyData worksheet displayed, display the Developer tab of the ribbon. (If no such tab is visible, click the Office button, click Excel Options, choose Popular at the left of the dialog box, and then select the check box labeled "Show Developer Tab in the Ribbon.")
4. Using the Insert tool in the Controls group, draw a Combo Box control somewhere on the MyData worksheet. (Make sure you use the Form Controls combo box, not the ActiveX Controls combo box.)

Follow these steps if you are using an older version of Excel:

1. Create a new worksheet and name it something like "ChartData".
2. Copy the column headers from the MyData worksheet to the second row on the ChartData sheet. (In other words, copy MyData!A1:P1 to ChartData!A2:P2. This leaves the first row of the ChartData sheet temporarily empty.)
3. With the MyData worksheet displayed, choose View | Toolbars | Forms. The Forms toolbar should be displayed.
4. Using the Forms toolbar, draw a Combo Box control somewhere on the MyData worksheet.

Regardless of which version of Excel you are using you now need to modify the properties associated with your combo box:

1. Display the Format Control dialog box for the newly created Combo Box. (Right-click the Combo Box and choose Format Control.)
2. Using the controls in the dialog box, specify the Input Range as **MyData!$A$2:$A$151**, specify the Cell Link as **ChartData!$A$1**, and specify the Drop Down Lines as 25 (or whatever figure you want).

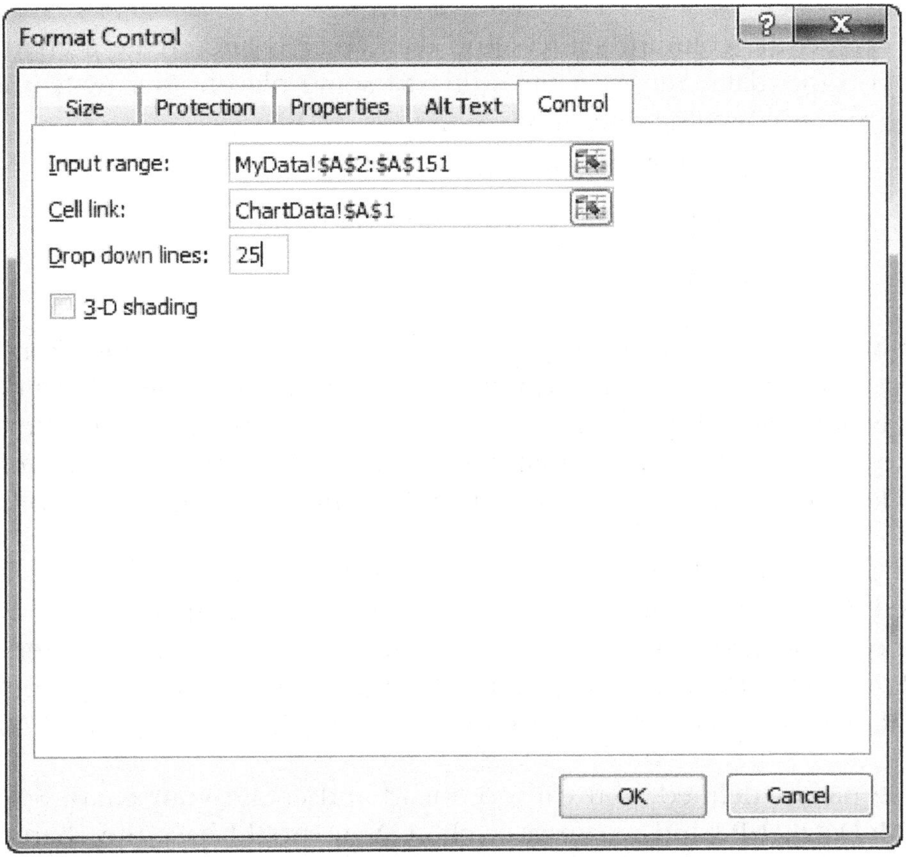

The Format Control dialog box.

3. Click OK to dismiss the dialog box. You now have a functioning Combo Box that, once you use it to select a company name, will place a value in cell A1 of the ChartData worksheet that indicates what you selected.
4. With the ChartData worksheet displayed, enter the following formula into cell A3:

```
=INDEX(MyData!A2:A151,$A$1)
```

5. Copy the contents of cell A3 to the range B3:P3. Row 3 now contains the data of whatever company is selected in the Combo Box.
6. In cell B1 enter the following formula. (The result of this formula will act as the title for your dynamic chart.)

```
="Data for " & A3
```

7. Select the column headers and data (B2:P3) and create a chart based on this data. Set the chart's title to some placeholder text; it doesn't matter what it is right now.
8. In the finished chart, select the chart title.
9. In the Formula bar, enter the following formula:

```
=ChartData!$A$3
```

You now have a fully functioning dynamic chart. You can use the Combo Box to select a different company, and the chart is redrawn using the data for the company you select. If you want, you can move or copy the Combo Box to the sheet containing your chart so that you can view the updated chart every time you make a selection. You can also, if desired, hide the ChartData worksheet.

A third approach is to use a macro to modify the range on which a chart is based. To prepare for this approach, create two named ranges in your workbook. The first name should be ChartTitle, and it should refer to the formula =OFFSET(MyData!$A$1,22,0,1,1). The second name should be called ChartXRange, and it should refer to the formula =OFFSET(MyData!$A$1,22,0,1,15).

With the names defined, you can go ahead and create your chart. Select the range MyData!B1:P2 and create your chart. You should base the chart on this simple range, and make sure that you place some temporary text in the chart title. Make sure the chart is created on its own sheet and that you name the sheet ChartSheet.

With the chart created, right-click the chart and choose Source Data. Excel displays the Source Data dialog box, and you should choose the Series tab.

Since you based the chart on the headers and a single row of data, there should be a single data series in the dialog box. Replace whatever is in the Values box with the following formula:

```
='Book1'!ChartXRange
```

Make sure you replace Book1 with the name of the workbook in which you are working. Click OK, and the chart is now based on the named range you specified earlier. You can now select the chart title and place the following in the Formula bar to make the title dynamic:

```
=MyData!ChartTitle
```

Now you are ready to add the macro that makes everything dynamic. Display the VBA Editor and add the following macro to the code window for the MyData worksheet. (Double-click the worksheet name in the Project Explorer area.)

```
Private Sub Worksheet_BeforeDoubleClick(ByVal Target As Excel.Range, Cancel
As Boolean)
    ActiveWorkbook.Names.Add Name:="ChartXRange", _
        RefersToR1C1:="=OFFSET(MyData!R1C1," & _
            ActiveCell.Row - 1 & ",1,1,15)"
    ActiveWorkbook.Names.Add Name:="ChartTitle", _
        RefersToR1C1:="=OFFSET(MyData!R1C1," & _
            ActiveCell.Row - 1 & ",0,1,1)"
    Sheets("ChartSheet").Activate
    Cancel = True
End Sub
```

Now you can display the MyData worksheet and double-click any row. (Well, double-click in column A for a row.) The macro then updates the named ranges so that they point to the row on which you double-clicked, and then displays the ChartSheet sheet. The chart (and title) are redrawn to reflect the data in the row on which you double-clicked.

Made in the USA
Monee, IL
08 November 2026

67816260R00079